Technological independence:
The Asian experience

Note to the reader from the UNU

The interface between science, technology, and society has been one of the major research concerns of the United Nations University since its inception. The Programme on "Technology Transfer, Transformation, and Development; The Japanese Experience," for instance, enquired into the infrastructure of technology, human resources development, and social and economic conditions, and analysed the problems of the technological development process in modern Japan.

The University decided to continue to examine this important subject in the light of different experiences in selected countries in Asia in a comparative, cross-cultural, and interdisciplinary framework. The present volume presents the findings of collaborative research by Asian scholars and scientists. The country research teams looked into the socio-economic factors for self-reliance, or its creation of a sustained growth process in science and technology for national development, with a view to providing policy makers, business leaders, and researchers with policy guidelines and research agenda based on empirical research.

Readers are also invited to refer to several publications emanating from another Programme on "Technological Prospective for Latin America," such as "Las nuevas tecnologías y el futura de América Latina" and "El futuro ecológico de un continente."

Technological independence: The Asian experience

Edited by Saneh Chamarik and
Susantha Goonatilake

**United Nations
University Press**

TOKYO • NEW YORK • PARIS

United Nations University Press
The United Nations University, 53-70, Jingumae 5-chome,
Shibuya-ku, Tokyo 150, Japan
Tel: (03)3499-2811   Fax: (03)3499-2811
Telex: J25442   Cable: UNATUNIV TOKYO

Typeset by Asco Trade Typesetting Limited, Hong Kong
Printed by Permanent Typesetting and Printing Co., Ltd.,
Hong Kong
Cover design by Apex Production, Hong Kong

UNUP-758
ISBN 92-808-0758-7
United Nations Sales No. E.93.III.A.1
04000 P

Contents

Introduction

The Asian region is home to several ancient sophisticated civilizations and cultures. Its population of over 3,100 million constitutes nearly 60 per cent of the world's total,[1] and its economies are a major source of growth in the world. The economy (GDP) of South and East Asia grew at a rate of 7.0 per cent for the period 1981–1990,[2] which was more than double the world average of 3.2 percent,[3] or that of the developed market economies (a category which also includes East Asian Japan), whose average growth rate was 3.0 per cent.[4] The countries of East Asia particularly are tilting the world's centre of economic and industrial gravity away from the North Atlantic area.

Other parts of the region, such as South Asia, have also achieved credible industrial growth, though this is as yet less spectacular than that of some of the East Asian ones. Thus, India, although having considerable problems, grew at an annual rate exceeding 3.6 per cent from 1965 to 1990,[5] and indeed for the period 1980–1990 it grew by 5.3 per cent.[6] This can be compared with the growth, say, of the first industrial nation, the United Kingdom, from 2 to 3 per cent in the nineteenth century.[7] And during the period 1965–1990 India's manufacturing capacity grew at 6–7 per cent a year.[8]

Thus, since the Second World War, the Asian region has shown remarkable growth in its industrial capacity. Many of these countries will, therefore, be key industrial players in the world economic arena. The period covered by the case-studies presented here is thus significant as a turning-point.

The role technology plays today in the development of Asian countries is dependent on many contingent factors. These include local history and social structures, traditions of technology, and local perceptions of the growth of industry in the West and the reactions that this engendered. A sketch of these contingent factors is an essential backdrop to an understanding of technology in contemporary Asia.

The per capita income of the region before the nineteenth century was not very different from that of Europe. The differences arose primarily after the Industrial Revolution. There is also some evidence that the success of the Industrial Revolution in Britain was paralleled by the partial destruction of manufacturing capacity in parts of Asia.[9]

The Asian region witnessed many scientific and technological developments before its encounter with the Europeans. These varied from knowledge of algebra and trigonometry in parts of South Asia to that of the compass and gunpowder in East Asia, and included many useful technological devices. In fact, the gradual development of science and technology in Europe was initially due in part to the transfer of some of these concepts and technologies to Europe and their interaction with the dynamic social, economic, and cultural forces let loose there after the Renaissance.[10] Before this period, there was also much intraregional traffic of ideas and knowledge *within* Asia.

The region and the global historical setting

The rapid development of technology in eighteenth- and nineteenth-century Europe had a positive feedback character, one technology development feeding another, leading to a spiral of technological development and economic growth. These occurred largely in the textile industry in Britain. In the cotton industry, innovations in spinning stimulated further improvements in other parts of the closely linked chain of cotton production. As a result, new forms of mechanization gradually emerged throughout the entire chain. Ultimately, this process of innovation led to the emergence of advanced machines that made inevitable – because the new technology was clearly far superior to the existing forms of manufacture – the development of large factories using these machines. The new technology helped turn cotton into a big industry, and as such its effect was felt throughout the land. This British experience became a model and an incentive for other countries to embark on the new technological road.[11]

The Industrial Revolution in Britain and the spurt in technology was preceded by the scientific revolution which reordered the formal perceptions of the natural world. Yet the technological spurt and the Industrial Revolution were not directly due to these new perceptions of the natural world brought about by the scientific revolution.[12] In fact, many of these developments were carried out by matter-of-fact technicians rather than scientists.

In the twentieth century, particularly since the Second World War, and more spectacularly in the last 20 years, technological developments have been directly linked to scientific advances. Many technological developments today, such as those in the front-line areas of biotechnology and information technology, are directly linked to research.

The technological growth that emerged in Europe was intimately linked to its economic development. This technological and economic development and the changed economic relations it brought about spread beyond Europe. These new global relations were associated with the search for raw materials for the new industries as well as for a partial market for the new industrial products. The relationships, however, were not limited to purely economic ones. A map of the world before the period on which we concentrate in this book would show large areas under the direct or indirect tutelage of Europe or its settler bastions.

The asymmetrical relationship of political power and economic power between Europe and Asia was also reflected in knowledge, technology, and culture, although the cultural intolerance which marked the mercantile era of expansion, with its attempts at mass religious conversion, had mellowed and the transfer of ideas had changed character in the industrial era. Transfers of knowledge, including sometimes those of industrial technology, were a part of the interaction between Europe and its colonies in the period before the Second World War.

This second phase of transfer of Western ideas can be traced to a fundamental change in geopolitical relationships from the nineteenth century on, with metropolitan economic interests centred largely on investment in mines, plantations and other activities that produced raw materials for the Industrial Revolution. This re-division of labour also went in parallel with the new skills needed for activities associated with the colonial presence. These requirements were often met by two processes: the transfer of institutions modelled on those of the centre and the transfer of an educational package up to university level.[13]

As the industrial era progressed, there were many significant changes in technology and industry, as well as in economic and political relations, between Europe and Asia. Asia and its perceived riches were the initial *raison d'être* of mercantilist voyages and subsequent

3

European expansion dating from the sixteenth century. In contrast to the situation in the Americas, or later in the nineteenth century in Africa south of the Sahara, the Asian encounter with Europe was not one of simple and easy subjugation. The Asian region's powers of resistance were much greater than those of the other two continents.

The countries nearer Europe, such as South Asia, gradually succumbed in varying degrees to European political and economic control, partly owing to the judicious use by outside powers of internal conflicts. The further east one went, the weaker were the communication and supply lines from Europe, and the more superficial was the penetration. However, China, though not completely colonized, suffered a partial internal collapse and was subject to a set of unequal treaties, with a consequent loss of sovereignty. At the extreme end of the European supply line was Japan, one major country apart from Thailand that defied European penetration after an initial encounter. It sealed off its borders after this initial contact till late in the nineteenth century, except for a nominal window in Nagasaki open to the Dutch.

The social changes engendered by the Asian interaction with Europe gave rise to a particular dynamics in the internal economic and manufacturing structures of the various Asian countries. In the centres of penetration of European commerce, local commercial groups grew up often as allies of the foreign power, as in cities like Calcutta and Shanghai. Some of them, like the *comprodores* of China, engaged in trade only as intermediaries, whilst others gave rise to the beginnings of a local industrial class, as in some parts of India.

The penetration and the partial subjugation of Asia by Europe gave rise to various local social responses aimed at both political and economic independence. Often, the debates surrounding the two went hand in hand. The debates themselves were often articulated by different strata who themselves were sometimes partial products of the European encounter.[14] Many of these debates began in the nineteenth century and continued in the twentieth, their effects having strong echoes in the era subsequent to the Second World War.

The Japanese response, after the initial decision to encounter the challenge from the West by adopting its technology and other characteristics, was perhaps one of the most single-minded outcomes of such a debate. Once it had been decided to open up the country after the Meiji Restoration, a policy of emulating and absorbing the technology and other socio-economic forms of the West was pursued with zeal. Japan was a uniform culture and absorption of Western influences

4

was, therefore, relatively easy. Furthermore, it had escaped subjugation, it did not have strata that were derived from a colonial presence, and its intellectuals did not possess a "wounded psyche." All these factors allowed for a very pragmatic and rapid absorption of Western-derived technology. Although this absorption was considered to be selective, as reflected in slogans such as *wakon yosai* (Japanese spirit, Western civilization), the technology was in fact absorbed largely unchanged. This was achieved essentially by developing the existing social, economic, and technological structures, the earlier system in many ways blending seamlessly into the new.[15] It was as though Japan, in its desire to possess particular industries, had gone shopping around the world with no psychological hang-ups and bought what it wanted.

Another country that escaped direct colonial penetration, albeit by a judicious geopolitical balancing act between various foreign powers, was Thailand. Thailand again possessed continuity of economic, technological, and cultural structure. This meant that when it decided upon industrialization, it could do so by "shopping around" for industrial development, basing its decisions on the perceptions of local social and economic structures.

Yet it is in the other countries of Asia, which were under varying degrees of political subjugation, that more "theoretical" (as opposed to pragmatic) approaches to industrialization and technology acquisition arose. In countries such as India and China, the search for industrialization went together with the search for national liberation. The discussions in these two major countries strongly influenced their neighbours.

A common thread in India and China was the example of the Soviet Union, which had apparently made a successful attempt to industrialize in the short time that had elapsed since the late 1920s, mainly as a result of its first two Five-Year Plans. Stalin's attempt at industrialization was widely admired at a time when neither the nature of that industrialization nor its human costs were yet visible.

Soviet society was also guided by a philosophy – Marxism-Leninism – which seemed to provide an explanation for the subjugation of Asian countries ("imperialism") as well as for their emancipation ("national liberation struggles"). The logic of liberation struggles in the original Marxian formulation was followed in China, whilst a home-grown emancipatory logic largely based on the teachings of Gandhi was followed in India. Each of these forms of emancipatory logic also had its explicit and implicit logic of production.

These different forms of logic were influenced by decision-making élites, existing socio-economic structures, and "the changing situation on the ground." In the case of India, a major element of the debate on relations with the West was Gandhi's rejection of urbanism and industrialism and his recommendation of a village-based economy. Yet the industrial policy that was finally adopted was that of Nehru. This was influenced strongly by the Soviet model and by Nehru's socialist orientation. The pro-industrialization policy was also helped by the many local industrialists who formed an important part of the power base of the Indian Congress Party. There was, however, a major difference from the Soviet model. Planned industrialization in India was being attempted within the framework of a market economy without a revolution, and without an omnipotent Politbureau. A further departure was that the vestiges of Gandhian policy still remained strong in manufacturing, in the protected handicrafts sector.

If the Indians veered away from the strict Soviet model, so too did the Chinese model. The Chinese, in waging a protracted guerrilla war primarily in rural areas, had to support themselves economically and so had to develop a viable production base.[16] This meant the adoption of policies that could help them meet minimal manufactured goods requirements. This practical exercise in industrial production before the communist take-over in 1949 was to colour the Chinese experience, leading it away from the strict, theoretically derived model of industrialization that was arrived at in the Soviet Union *after* the revolution. This specifically Chinese tendency was intensified in the later years, particularly when the Soviet Union withdrew its initial support after the Sino-Soviet conflict.

If the example of the Soviet Union was a strong influence on China, communist South-East Asia, and South Asia, countries in East Asia had different geopolitical influences. The Republic of Korea, Taiwan, and Hong Kong were ranged on the opposite side of the ideological spectrum in that they were all strongly anti-communist states, both having as adversaries "parallel entities" in the form of mainland China and the Democratic People's Republic of Korea. They pictured themselves as strong ideological and economic adversaries of communism. They benefited strongly from a nurturing and aid relationship with the West, primarily the US and, to a lesser extent, Japan and the other OECD countries. Yet these economies, although governed by market principles, contained an implicit admixture of the Japanese and Soviet experiences, both of which were led by strong central government guidance. This meant that in a country like the Republic of Korea, the impact of the state was initially very strong in key industries.

If the Republic of Korea and Taiwan were the direct adversaries of their "parallel entities" in the communist bloc, several other countries in South-East Asia were drawn into a strong anti-communist stance because of internal political and geopolitical considerations. These countries included Hong Kong, Singapore, Thailand, Malaysia, and Indonesia. The economic policies followed by them included – except in the case of Hong Kong – state guidance of the economy combined with strong economic support from Western countries, primarily the US, and a distinct bias towards the private sector.

We have described the geopolitical background at the time the countries of Asia began consciously to implement policies of industrialization. It was a dynamic background, with several of the external and internal givens changing rapidly after the onset of industrialization. Some of these internal dynamics were the result of social and economic changes brought about partly by the success or failure of the initial economic and industrial policies. A significant external factor was the changing world economic and technological environment.

Post-Second World War geopolitics

The structure of production and international trade at the end of the Second World War closely fitted a colonial mould. The key metropolitan countries in Europe and America were closely linked with their various subjugated lands in Asia. This meant that the bulk of the trade with these Asian countries and the West centred on the export of primary produce and the import of finished goods to Asia. Colony or near-colony was closely linked economically to the "mother" country. Little trade from other European countries penetrated the colony. This was the pattern established in countries like Korea, which was colonized by Japan, whilst in Thailand, which was less directly linked to a single European country, a greater spread of imports and exports was possible. As colonialism declined in the 1940s and 1950s, trade patterns changed.

This transformation was also helped by the intensified trade within the "mother" countries themselves, as post-Second World War re-arrangements also led to a greater expansion of intra-European trade and the lowering of economic barriers. The world was in some ways becoming less parochial and forging new connections; in other ways it was becoming more parochial, with an intensification of intraregional links.

The decrease in parochialism is seen in the globalizing tendencies which have become a central feature of the modern world. Commerce, trade, ideas, and culture travel around the world much faster and in a more thoroughgoing way than before. Because of widespread travel possibilities and exploding global telecommunication links, distance becomes less and less important and the world is increasingly inter-linked.[17] No country in this deepening process can remain an island, and no economy a single player.

Economic, technological, or cultural autarchy is not possible in this interdependent world, as the implosion of Eastern Europe vividly demonstrates. Yet, as the world became increasingly interconnected, it became clear that all the interconnected players were not equal. A dynamic new web of economic and other relationships between Asia and the rest of the world replaced the relative rigidities of the colonial world. These relationships, although often perpetuating older unequal relationships, through their very dynamism permitted new configurations and new opportunities, encouraging the mastery of external technology, the acquisition of internal industrial capacity, and the emergence of new patterns of trade and commerce within as well as between countries.

The countries of the region responded in varying ways to the changing circumstances of the ex-colonial world. Countries such as India and China, and those influenced by their respective strategies, followed the path of import substitution and even of near-autarchy. Other countries, such as the Republic of Korea, initially followed import-substitution strategies and later became export-oriented. City-states like the ports of Hong Kong and Singapore had more open policies towards the external world, though Singapore's were more government-directed than the *laissez-faire* policies of Hong Kong. Other countries fell somewhere in between the openness of Hong Kong and the more inward-looking policies of India and China.

As independence receded into the past, the limits of import substitution, even for the larger autarchial-oriented countries, was increasingly felt. The subsequent process of lowering the economic and technological barriers to the outside world was also influenced by external financial and lending agencies, such as the IMF, the World Bank, and regional banks, as well as by globalizing tendencies that bound multinational companies deeply into the global economic fabric. If the earlier multinational companies that penetrated Asia in the colonial period were associated with one parent nation, the new multinationals were increasingly broad-based and dynamic. More and more, their ownership itself was traded publicly and across borders. The search for profit maximization of these large economic entities led to symbiotic relationships with smaller economic entities in Asia, relationships that covered not only trade but also technology and manufacturing capacity. Some industries from the metropolis were relocated in Asia, initially because of the rise in labour costs in the metropolitan countries.

Early commentators on this redivision of labour emphasized only the search for cheap labour, but ignored or minimized the technology transfer that this entailed. More recent studies of such economic relationships, especially in South-East Asia, have suggested that the search for profit maximization of MNCs also led to the concurrent development of technological capacities in the Asian region, including the emergence of significant R&D facilities.[18] Capital, technology, and, to a lesser extent, labour had become much more mobile in the new emerging global order. In 40 years, the colonial order had been transformed into a more vital and dynamic entity.

This continuing globalization was helped not only by geopolitical considerations, such as the break-up of the colonial system, but also by key changes in technologies.

New technologies

When industrialization began in the late eighteenth and early nineteenth centuries, technology associated with the steam engine was a key factor. Since then, successive technologies have transformed manufacturing and consumption, and these waves affected different countries at different times. Those countries that began to industrialize in the nineteenth century, for example, first incorporated the earliest industrial technologies and later added other generic technologies as they emerged. In this, the early latecomers to industrialization followed the historical sequence of technology development in the West, although in a telescoped form.

Those that came later still – particularly those that embarked on major industrialization after the Second World War – encountered a world of generic technology that had a greater spread than existed in the early years of industrialization. At the heart of our description of the mastery of technology in Asia in the last 40 years is the differential importance and impact of the generic technologies.

The key technologies in order of emergence are: steam; electricity; chemicals, oil-based chemicals, and synthetic materials; and, in the contemporary period, information technology and biotechnology.

One should note here that new technologies, as they develop, pass through the economy in "creative waves of destruction," in the words of Schumpeter,[19] destroying the old and establishing the new. This process gives rise to a new range of products together with new technical means of manufacture. Steam was one such technology, powering the early systems of mass manufacture. Its penetration, in the form of application to products and processes, achieved a plateau in the latter half of the nineteenth century, around the time that Industrial Revolution technology was beginning to penetrate parts of Asia.

The next technology was associated with electricity.[20] This gave rise to a wide range of applications, changing not only the motive power in manufacture from steam to electricity but also making a qualitative shift. Here, the motive power associated with the electric motor signalled a change in the organization of the factory, from one with a centrally placed power source based on a steam engine, with a clutter of belts to transmit the power, to a more decentralized system with several individual manufacturing operations powered by individual motors. The new electrical technology also gave birth to a variety of new products in the new forms of light, heat, and motive power. This "wave" rose rapidly in the latter half of the nineteenth century and reached a plateau in its applications in the West around the 1950s, at a time when the bulk of the Asian region was launching itself on a concerted industrial programme.

In similar fashion, the technologies associated with the internal combustion engine, oil-based chemicals, and synthetic materials emerged around the beginning of this century, and their applications then increased rapidly,[21] levelling out around the 1970s, at a time when most countries in Asia were beginning a shift from import-substitution and quasi-autarchic policies to a more global orientation.

The two new technologies that are rapidly maturing at present, namely information technology and biotechnology, are expected to have a much more pervasive impact than the earlier technologies. Information technology will replace many human functions on a mental level, in the some way that the steam engine replaced human and animal motive power.[22] A wide variety of hitherto human roles, from the highly skilled to the most routine, are being penetrated by the new technology. In addition, this technology is able to perform new mental functions formerly beyond the scope of humans. It will therefore penetrate not only manufacturing systems and manufactured products but other sectors of the economy as well, including many human functions in such spheres as administration, marketing, and finance. It should be noted that the number of applications of the new information technology is rising rapidly, much more rapidly than did the earlier technologies, and a levelling-off – if there is to be one – is not in sight.

The other new technology, biotechnology, does not replace human skills but it gives rise to a very wide-ranging array of products as well as manufacturing processes. Its range of applications is increasing rapidly as new uses are discovered in fields from health to agriculture and manufacturing. The degree of penetration of this technology is at present unclear. Three possible trajectories for growth have been surmised by Freeman,[23] but since his projection appeared in 1987 development in this field has accelerated, and it now seems that the faster and more pervasive scenario for the new technology is the more likely.

These two new technologies, which began to make an impact in the 1970s and 1980s, are arriving on the Asian scene at a time when Japan has caught up with the West in industrial terms and when several other Asian countries have passed their import-substitution phase and are beginning to open up to the outside world.

It is also important to note the interplay between the different forms of technology and the time dimension in the Asian countries. The different generic technologies that we have described have been presented to the different countries in the region at different times, at different points in the maturation of their economies, and at different periods in the global socio-economic system. In general this meant that, the later the technological arrival, the greater the spread of generic technology that one could shop for and the more telescoped the technological history that had to be acquired and mastered.

Drawing the various threads together, it is clear that the context within which the mastery of technology occurs in the Asian region in the period under review consists of several factors. These include the global geopolitical structures subsequent to the Second World War and their later rearrangement, particularly *vis-à-vis* relations with the West where modern technology first appeared. Many of the Asian countries emerged as political entities after momentous internal upheavals, revolutions, independence struggles, and wars; and, after they had taken their different paths towards industrialization, the global political and economic environment, as well as the nature of technology itself, continued to change. These different factors played themselves out and interacted dynamically with each other as the countries attempted technology acquisition. It is the resulting industrial directions, and their associated problems, that are charted in this research study.

The study

The study was undertaken by several national teams in China, India, Japan, the Republic of Korea, the Philippines, and Thailand. The study covered two main aspects: first, a description of the internal dynamics of policy and socio-economic factors under which technology acquisition was carried out, and, second, illustrative case-studies of selected key industries. In the chapters to follow, different countries' broad national-level experiences in technology mastery will be discussed from the point of view of self-reliance. For reasons of space and presentation, the details of case-studies will be omitted.

The study was coordinated by Saneh Chamarik and carried out by the country teams whose names are given at the end of the book. The original research reports upon which the country chapters are based were edited by Susantha Goonatilake; the chapters were subsequently vetted for fidelity to the original by the country teams. The introduction and the final chapter were also written by Susantha Goonatilake.

Notes

1. *World Population Prospects*, New York: United Nations, 1990, p. 22.
2. *World Economic Survey*, New York: United Nations, 1990, p. 3.
3. See note 2 above, p. 1.
4. See note 2 above, p. 1.
5. *Trends in Developing Economies*, Washington, D.C.: World Bank, 1991, p. 275.
6. *World Development Report*, Washington, D.C.: World Bank, 1991, p. 206.
7. Phyllis Deane, *The First Industiral Revolution*, Cambridge: Cambridge University Press, 1979, p. 291.
8. See chapter 1 of this volume.
9. Lucille Brockway, *Science and Colonial Expansion*, New York: Academic Press, 1979, p. 24.
10. Donald F. Lach, *Asia in the Making of Europe*, vol. II, Chicago: University of Chicago Press, 1977.
11. D.S. Landes, *The Unbound Prometheus*, London: Cambridge University Press, 1970.
12. J.D. Bernal, *Science in History*, New York: Hawthorn Books Inc., 1956.
13. K.M. Panikkar, *Asia and Western Dominance*, London: George Allen & Unwin Ltd., 1953, p. 317; Eric Ashby, *Universities British, Indian, African: A Study in the Ecology of Higher Education*, London: Weidenfeld & Nicolson, 1966, pp. 63, 111; B.V. Subarayappa, *A Concise History of Science in India*, New Delhi: National Science Academy, 1971, p. 500.
14. See, for example, V.C. Joshi, ed., *Ram Mohun Roy and the Process of Modernization in India*, New Delhi: Vikas, 1975.
15. M.Y. Yoshino, *Japan's Managerial System – Tradition and Innovation*, Cambridge, Mass.: MIT, 1968.

16. Edgar Snow, *Edgar Snow's China: A Personal Account of the Chinese Revolution*, New York: Random House, 1981.
17. Joseph N. Pelton, "Global Talk and the World of Telecommuterenergetics," in Howard F. Didsbury, ed., *In Communication and the Future: Prospects, Promises and Problems*, Bethesda, Md.: World Future Society, 1982.
18. J. Henderson, *The Globalisation of High Technology Production*, London and New York: Routledge, 1989.
19. J. Schumpeter, *Business Cycles*, New York: McGraw Hill, 1939.
20. OECD, *Biotechnology: Economic and Wider Impacts*, Paris: OECD, 1989, p. 54.
21. See note 20 above.
22. OECD, *New Technologies in the 1990s: A Socioeconomic Strategy*, Paris: OECD, 1988.
23. See note 22 above.

1

India

Background

Post-independence Indian policy makers recognized early the role of science and technology (S&T) self-reliance in endogenous economic development. In fact, a Scientific Policy Resolution was adopted by the Indian Parliament as early as 1958, and the basic premise of the second Five-Year Plan (1956–1961) was that the acquisition of capital goods-making capability was vital for long-term self-reliant economic development. Since then, Indian planning and policies have gradually evolved to develop local S&T capabilities.

The present analysis has been conducted in two stages. First, technological policies and their outcomes over the post-independence period are studied within the overall development framework. Next, case-studies are presented for four manufacturing industries: machine tools, coal-based thermal power equipment, petroleum refining, and chemical fertilizers. Each of these industries is intensive in its use of scarce resources, namely capital, skills, and technology, and has many linkages with other sectors.

In this study, S&T self-reliance refers to the ability to apply knowledge for national development, making no distinction between the two constituent elements, science and technology. It is used here to mean not autarchy or self-sufficiency in technology but an autonomous capability to apply technology – whether local or imported – for national development.

Development perspectives in the Indian economy

India's development effort, beginning with the first Plan, has emphasized raising the domestic savings and investment rate in order to achieve higher growth and faster industrialization. The second Plan provided the requisite elements of industrialization for this strategy, stressing the increased domestic manufacture of capital goods through the development of "basic" industries.

Given India's ample natural resources, it was recognized that the key to industrialization lay, first, in establishing a manufacturing capacity in heavy machinery, heavy electrical equipment and machine tools. As this machinery became available it would be possible, gradually, to manufacture everything else. The next four stages of the logic of development were: steel to make machinery and electricity to drive the machinery; engineers, technologists, technicians and skilled workers to convert raw material resources into machinery and power; the expansion of applied S&T research to solve practical development problems and also to expand the S&T knowledge stock through fundamental or basic research; and a sufficient number of persons with S&T capabilities to undertake the above tasks. Increasing the supply of S&T manpower came to be considered "the only secret" of fast development for a big country like India.[1]

The above logic of development is reflected in the structural changes envisaged in the second and subsequent Five-Year Plans in favour of relatively more growth in the basic and capital goods industry, emphasis on import substitution in steel, fertilizers and oil, the high priority given to higher and technical education, and an increasing allocation of resources to R&D in government institutions.

The strategy aimed at faster growth and self-reliance was characterized by an investment pattern in favour of the development of basic industries and physical infrastructure. The public sector played a significant role in their development, increasing the allocation to higher and technical education and the development of central institutions; this was accompanied by a host of policies and measures for the promotion of indigenous industry and technology, the growth of domestic savings and investment, the diversification of industry and trade in favour of manufactures and high value-adding products, and self-sufficiency in essential items like food. The basic thrust of the strategy was the acceleration of the process of domestic capital and technological accumulation. It was also oriented to strategic and security considerations.

15

The economic progress of India over the last four decades is in many ways impressive. The long-term growth rate is now well over 4 per cent annually, compared to only 1 per cent in the decade before independence. Since independence, India's real Gross National Product (at factor cost and at 1970/71 prices) increased from Rs. 178,410 million in 1951/52 to Rs. 612,010 million in 1984/85, in a period covered by six Five-Year Plans and three annual plans. This was also accompanied by many important structural changes in resource mobilization, and in the patterns of value added and foreign trade.

The process of structural change was a result of the accelerated pace of capital accumulation, which had more than doubled over the last 30 years. The rate of aggregate gross investment (domestic capital formation), which was only 11.9 per cent in 1951/52, had increased to 23.4 per cent in 1984/85, the public and private sector contributing almost equal shares. It is important to note that the bulk of this increased investment was made through domestic (gross) savings, which increased from 10.0 per cent in 1951/52 to 22.1 per cent in 1984/85, as table 1 shows. Net foreign aid contributed as little as 4 per cent of aggregate investment under the recent Five-Year Plans.[2] Thus, significant progress in financial self-reliance has been achieved since the beginning of the planning period. The projected contribution of foreign inflow in the comprehensive seventh Plan (1985–1990) is little more than 5 per cent, although recently import policy has been liberalized in a significant way. Many other indicators, such as the marginal role of foreign private investment, suggest that India's performance in respect to "self-reliant development" compares very favourably with that of other developing countries.[3]

The manufacturing sector was the most dynamic sector, recording a growth rate of 6 to 7 per cent during the review period and increasing its share of the GNP to over 15 per cent in 1984/85.[4] The manufacturing sector was not only growing faster but also growing in complexity, in that the share of basic and capital goods industries in the gross value added of the manufacturing industries increased substantially to over 68 per cent, at 1960 prices, in the period 1960–1980 (table 2). The share of technology-linked – i.e. capital goods – industries increased from 20.3 to 27.4 per cent over this period, and that of the electrical and non-electrical machinery industries from 9.0 to 16.8 per cent (table 3). This emerging industrial structure is highly conducive to S&T development.

Furthermore, there was substantial growth in, and structural transformation of, India's foreign trade sector over this period. Exports

Table 1. Net foreign aid utilized by India: share in aggregate (gross) and public sector investment (capital formation) (Rs. million at current prices)

Period (1)	Net aid utilized[a] (2)	Investment Aggregate[b] (3)	Investment Public sector (4)	% share (2) in (3)	% share (2) in (4)
First Plan (actuals), 1951/52 to 1955/56	1,780	54,080	19,600	3.3	9.1
Second Plan (actuals), 1956/57 to 1960/61	13,110	101,270	46,720	12.9	28.1
1960/61	2,622[c]	25,440	11,420	10.3	23.0
Third Plan (actuals), 1961/62 to 1965/66	23,250	167,450	85,770	13.9	27.1
1965/66	8,410	43,900	23,323	19.2	36.1
Annual Plans (actuals), 1966/67 to 1968/69	22,470	158,840	66,250	14.1	33.9
1970/71	3,410	73,440	27,730	4.6	12.3
Fourth Plan (actuals), 1969/70 to 1973/74	17,390	414,020	155,790	4.2	11.2
1975/76	11,540	148,110	64,169	7.8	18.0
Fifth Plan (actuals), 1974/75 to 1978/79	35,390	879,500	394,260	4.0	9.0
Sixth Plan, 1980/81 to 1984/85 (projections at 1979/80 prices)	58,890	1,587,100	975,000	3.7	6.0
Sixth Plan (actuals)					
1979/80 (base year)	5,520	252,830[d]	118,160[d]	2.1	4.7
1980/81	13,580[e]	311,850[d]	139,260[d]	4.4	9.8
1981/82	10,210	364,850[d]	175,280[d]	3.7	5.8
1982/83	13,020	398,110[d]	200,470[d]	3.1	6.2
1983/84	12,350	453,480[d]	217,730[d]	2.7	5.7
1984/85	11,780	497,810[f]	267,720[f]	2.4	4.4
Total, 1980–1985	60,940	2,026,100	1,000,460	5.6	6.1
Seventh Plan, 1985/86 to 1989/90 (projections at 1984/85 prices)	180,000 (200,000)[g]	3,223,660	1,542,180	5.6 (6.2)	11.7

Source: Ministry of Finance, *Economic Survey*, various issues; Central Statistical Organization (CSO), *National Accounts Statistics*, various issues; Planning Commission, *Sixth Five Year Plan, 1980–85*, 1981.

a. Data in column 2 do not include suppliers' credits and commercial borrowings.
b. Adjusted for errors and omissions.
c. Annual average, actual not readily available.
d. Provisional.
e. Includes drawings under IMF Trust Fund.
f. Quick estimate.
g. Parentheses point to the projections, including Rs. 20,000 of private corporate borrowings.

17

Table 2. Changing significance of capital goods industries: share of different industry groups in total[a] value added (gross, at 1960 prices), 1960, 1970, and 1980

Industry groups	1960	1965	1970	1975	1980	Point variation 1980/1960
Basic industries	24.95	34.45	38.33	39.39	40.45	15.50
Capital goods industries	20.34	24.29	24.33	25.73	27.43	7.90
Intermediate industries	5.05	3.79	4.92	3.96	3.50	(−)1.55
Consumer goods industries	49.06	37.47	30.42	30.92	28.62	(−)20.44
Total	100.00	100.00	100.00	100.00	100.00	−

Source: CSO, "Principal Characteristics of Selected Industries in Organised Manufacturing Sector, 1960 to 1980," December 1980 (mimeo).

a. Total of all selected industries in Annual Survey of Industries.

Table 3. Direction of change in capital goods industries: share of different capital goods industries in total[a] value added (gross, at 1960 prices)

Capital goods industry	1960	1965	1970	1975	1980	Point variation 1980/1960
Machinery except electric	4.23	6.54	6.95	7.40	7.89	3.66
Electric machinery, etc.	3.75	4.99	6.73	7.69	8.95	5.20
Shipbuilding and repairs	0.86	0.51	0.45	0.60	0.69	(−)1.17
Railroad equipment	3.96	3.22	2.33	1.63	2.12	(−)1.84
Motor vehicles	3.28	4.27	3.49	4.37	3.81	0.53
Metal products	3.05	3.52	3.04	2.57	2.69	(−)0.36
All (including repair of motor vehicles)	20.34	24.29	24.33	25.73	27.43	7.09

Source: CSO, "Principal Characteristics of Selected Industries in Organised Manufacturing Sector, 1960 to 1980," December 1980 (mimeo).

a. Total of all selected industries in Annual Survey of Industries.

grew from Rs.7,329 million to Rs.115,548 million between 1951/52 and 1984/85. This increase was due to diversification in product composition as well as in destination. The share of the manufacturing sector in total exports increased by over 50 per cent. The growth of manufactured exports contributed substantially to the growth of the manufacturing sector and to the increasing "economic openness" of the economy.[5] Similarly, there was change in the import basket in favour of material and capital inputs. Another development in imports has been the significantly reduced dependence on strategic products such as food, fibres, fuel, fertilizers, steel, and, indeed, machinery (table 4).

From the viewpoint of technological self-reliance, the most significant substitution was that of imports of machinery and equipment by domestic production. Considered in relation to capital formation in the form of machinery and equipment, import dependence fell from about 70 per cent in 1950/51 and over 60 per cent in 1960/61 to about 20 per cent by the end of the 1970s, and again to about 10 per cent in recent years; this works out to only 6.7 per cent in 1982/83, the last year for which data could be obtained on a comparable basis.

It should be noted, however, that in spite of these notable achievements, the performance of the Indian economy has fallen short of its planned targets as well as of its real potential. For instance, in every Plan except the last, performance has not matched the targets, in terms of either overall growth or the individual growth of major sectors. Nevertheless, many similarly placed developing countries recorded much slower growth rates and a higher dependence on the rest of the world than did the Indian economy.

There is growing evidence that the productivity of many sectors in India is low compared not only with the major developed countries but also with many developing ones. Thus, the increase in the incremental capital-output ratio (ICOR) over the Plan periods shows a decline in the capital productivity ratio.[6] There is also evidence to suggest that factor productivity grew very marginally in the Indian economy.[7] The growth of labour productivity (value added per employee) in industries is estimated to be around 1.5 per cent during the period 1965–1980. The growth was slower in basic and intermediate industries and negative in consumer goods industries, but higher in capital goods industries. However, this indicator varied considerably over the period and the overall trend may, therefore, not be statistically significant.

The energy-GNP intensity in the economy has also increased. Growing pressures on India's balance of payments, particularly a slower

Table 4. Import substitution in strategic products in India (percentage share of imports in total supplies/availability)

Product	Pre-Plan, 1950/51	End of 1st Plan, 55/56	2nd Plan, 60/61	3rd Plan, 65/66	Annual Plans, 68/69	4th Plan, 73/74	Variation col. 7/ col. 2	End of 5th Plan, 78/79	6th Plan period 80/81	81/82	82/83	83/84	84/85	Point variation, col. 14/ col. 2	7th Plan, 1989/90 (projection)
1	2	3	4	5	6	7	8	9	10	11	12	13	14	15	16
Food grains	5.9	1.7	4.7	9.5	5.6	4.3	−1.6	(−)a	(−)	0.6	1.4	3.5	1.8	−4.1	−
Steel	25.2	39.9	35.7	16.7	9.3	18.5	−6.7	14.8	16.7	14.2	27.2	25.3	17.4	−7.8	11.2
Machinery^b	68.9	41.0	40.7	27.8	24.6	17.0	−51.9	10.2	12.9	10.2	8.0	–	–	−60.9c	NA
Petroleum^d	92.5	93.8	94.6	76.6	66.2	70.8	−21.7	56.4	63.0	48.0d	38.0	37.3	29.5	−63.0	28.1–29.4
Nitrogenous fertilizers	72.5c	39.8	80.3	58.3	60.9	38.3	−34.2	36.0	41.1	24.4	11.1	15.8	33.9	−38.6	27.9–29.5

Source: Planning Commission, *Sixth Five Year Plan, 1980–85*, 1981, p. 15; *Seventh Five Year Plan, 1985–90*, 1985, chap. 3; Ministry of Finance, *Economic Survey*, various volumes.

a. Means net exports.
b. Imports as a percentage of machinery component of gross investment on calendar year basis.
c. Col. 12/col. 2.
d. Net imports in throughput (financial year).
e. For 1951/52.

growth of exports in recent years, is also a reflection of the loss of international competitiveness over the period.

It is not surprising, therefore, that in the seventh Five-Year Plan (1985–1990), the major concern of Indian planners was to focus on measures for raising productivity and on the reformulation of a strategy to improve the international competitiveness of Indian industry.

Since the second Five-Year Plan, policy emphasis in India has been on import substitution and industrialization based on heavy industries. The government's various policy instruments, notably trade policy, fiscal policy, industrial licensing policy, and technology policy, were aimed in this direction, in both the public and private sectors. In the phase of industrialization, the public sector played a dynamic role, taking considerable initiative in setting up new heavy industries and infrastructure.

The development of technology through investment in physical and human capital was a key component of the planning strategy; technological progress was to engineer growth and lead to social transformation.

The Indian economy has shown a significant increase in "economic openness" in recent years compared to the beginning of the 1970s, as indicated by the increased share of foreign transactions in GDP, collaborations, and investment.[8] The seventh Plan (1985–1990) lays special emphasis on the development of human resources through programmes of education, health, social welfare, and S&T. It recognizes the importance of "the technology revolution and the growth of human capital and communication" for the "advance technology which can . . . generate resources for accelerated growth."[9]

S&T in development plans and policies

The initial plans emphasized the creation of an institutional infrastructure in S&T. The aims of government policy for the development of S&T in India were set forth in the Scientific Policy Resolution of March 1958, which reflected the government's emphasis on the pure, applied, and educational aspects of S&T, on creating conditions that would lead to an increase in the supply of quality scientists, and on ensuring that the benefits derived from the acquisition and application of scientific knowledge would be enjoyed by the people.

The third Plan (1961–1966) indicated that, as a result of developments during the first two Plans (1951–1956 and 1956–1961), an extensive network of institutions engaged in scientific research had come

Table 5. Five-Year Plan outlays (public sector) in India: aggregate and S&T sector (Rs. million at base year or current prices)

Plan	Outlay allocated to S&T in public sector	Aggregate public sector outlay	% of S&T in aggregate
First Plan, 1951/52 to 1955/56	140	19,600	0.7
Second Plan, 1956/57 to 1960/61	330	48,000	0.7
Third Plan, 1961/62 to 1965/66	715	80,990	0.9
Fourth Plan, 1969/70 to 1973/74	1,423	159,000	0.9
Fifth Plan, 1974/75 to 1978/79			
Draft	4,190	372,500	1.1
Final	7,676	392,875	2.0
Sixth Plan, 1980/81 to 1984/85			
(actuals)	8,652	975,000	0.9
1980/81	974	148,324	0.7
1981/82	1,483	182,109	0.8
1982/83	2,081	212,829	1.0
1983/84	2,285	250,875	0.9
1984/85	3,375	302,321	1.1
Total	10,198	1,096,458	0.9
Seventh Plan, 1985/86 to 1989/90	24,660	1,800,000	1.4
1985/86 Plan	4,448	322,386	1.4

Source: Planning Commission, *Five Year Plans*.

into existence at a large number of centres. The third Plan allocated about 1 per cent of total public sector outlays for S&T (table 5).

In line with the development strategy of the second Plan, in the third Plan the priorities of "self-reliance" were stressed primarily in terms of concentrating on the expansion of the capital goods and machine-building industries, together with the corresponding development of mining, power and transport, on a scale that would enable the country to build up sufficient capacity to produce domestically the bulk of the capital goods and machinery it required for high investment.[10] In emphasizing "self-reliance" in the capital goods sector, the planners' aim was to apply this to several dimensions – material, financial, and technical. The third Plan also laid stress on "speedy and extensive utilization of scientific research."

The thrust of the fourth Plan (1969–1974) was on finalizing the time-scale for the achievement of "self-reliance," in terms of reducing net foreign aid to half of the then current level by the end of the Plan and eliminating it soon thereafter. This Plan, like the Third Plan, also allocated about 1 per cent of aggregate public sector outlay to S&T.

The draft fifth Plan (1974–1979) envisaged a comprehensive S&T Plan, covering all sectors of the economy. The National Committee on S&T (NCST) set up by the government was to undertake the task of formulating the programme context of the Plan and to advise on the policy framework, covering both S&T and research, design, and development (RDD).

The draft fifth Plan included several programmes aiming at technological growth in selected fields and envisaged a public sector outlay of Rs.4,190 million (1.1 per cent of total), compared to Rs.1,423 million in the fourth Plan (0.9 per cent of total). The total Plan and non-Plan outlay envisaged was Rs.15,682 million, compared to Rs.3,736 million in the previous Plan, amounting to a quantum jump of over 300 per cent. The final fifth Plan revised the total outlays upward. The outlay allocation for S&T was also raised (to Rs.7,676 million) in revision and was around 2 per cent of the total.

The development plans also emphasized the expansion of the base of higher learning and technical education. The public sector played the major role in this. The substantial public sector expenditure devoted to university education and other educational programmes can be inferred from its percentage of total expenditure on education (table 6), which increased from 18 in the first Plan to 27 and 26 in the second and third plans respectively, and to over one-third in the successive Five-Year Plans. In technical education, the percentage increased from 14 in the first to 18 in the second and 21 in the third plans. A large institutional base and a substantial stock of technical manpower was built up during this period. In the successive Five-Year Plans the share was 11–13 per cent, increasing from Rs.1,060 million in the fourth Plan to Rs.2,780 million in the sixth Plan.

As a result, there was a phenomenal increase in the number of technical and science degree-holding graduates and postgraduates in all sciences, and in agriculture, engineering, and medicine (tables 7 and 8). The stock of S&T (economically active) manpower (table 9) increased in aggregate from 188,000 in 1950 to 450,000 in 1960,

Table 6. Plan expenditure/outlay on education under the Plans (Rs. million, rounded at current/base year prices)

Head	1st Plan, 1951/52 to 1955/56	% of total	2nd Plan, 1956/57 to 1960/61	% of total	3rd Plan, 1961/62 to 1965/66	% of total	Annual Plans, 1966/67 to 1968/69	% of total	4th Plan, 1969/70 to 1973/74	% of total	5th Plan, 1974/75 to 1977/78	% of total	6th Plan, (outlay), 1980/81 to 1984/85	% of total
General education														
Elementary education	850	55	950	35	2,010	34	750	23	2,390	30	2,920	32	9,050	36
Secondary education	200	13	510	19	1,030	18	530	16	1,400	18	1,660	18	4,200	17
University education	140	9	480	18	870	15	770	24	1,950	25	2,040	22	4,860	19
Subtotal	1,190	77	1,940	72	3,910	67	2,050	63	5,740	73	6,620	72	18,110	72
Other educational programmes	140	9	270	9	660	11	330	10	940	12	1,130	12	3,510	14
Cultural programmes	—a	—	30	1	70	1	40	1	120	2	270	3	840	3
Total general education	1,330	86	2,240	82	4,640	79	2,410	75	6,800	87	8,020	88	22,460	89
Technical education	200	14	490	18	1,250	21	810	25	1,060	13	1,090	12	2,780	11
Total education	1,530	100	2,730	100	5,890	100	3,220	100	7,860	100	9,110	100	25,240	100
Expenditure on education as a percentage of total Plan expenditure	7.8		5.8		6.9		4.9		5.0		3.2		2.6	

Source: Ministry of Education and Culture, *A Handbook of Educational and Allied Statistics*; Planning Commission, *Sixth Five Year Plan, 1980–85*.

a. Included under other categories of general education.

Table 7. India: Number of degrees awarded annually in science and technology/engineering since 1950

	Graduates				S&T postgraduates						
Year (1)	BSc (2)	BSc Agri. (3)	BE/BSc Engi. (4)	Total (2 to 4) (5)	MSc, incl. Home Science (6)	MSc Agri. (7)	MBBS (8)	MD (9)	MS (10)	ME/MSc Engi. (11)	Total (6 to 11)
1950	9,054	1,000	1,358	11,412	984	94	1,725	58	30	–	2,891
1951	10,407	1,041	1,494	12,942	1,398	143	1,555	62	40	–	3,198
1952	10,208	870	2,018	13,105	1,641	156	1,864	73	40	1	3,775
1953	11,292	879	2,326	14,497	1,780	168	1,782	66	52	6	3,754
1954	13,390	910	2,776	17,076	2,146	151	2,518	56	54	4	4,929
1955	14,761	905	2,789	18,455	2,348	141	2,564	88	77	10	5,228
1956	14,947	893	3,432	19,272	2,529	152	2,434	87	84	7	5,293
1957	16,706	1,128	3,443	21,277	2,933	217	2,764	101	101	40	6,156
1958	17,469	1,520	3,720	22,709	2,942	221	2,856	151	130	75	6,375
1959	19,091	1,950	4,160	25,201	3,508	264	2,939	172	195	180	7,258
1960	21,238	1,990	4,899	28,127	3,642	422	3,181	206	191	135	7,777
1961	24,797	2,608	5,965	33,370	4,727	496	3,565	210	206	411	9,615
1962	25,563	2,609	6,459	34,631	5,195	576	3,567	285	240	302	10,165
1963	33,318	4,112	8,286	45,716	5,848	634	3,936	313	308	301	11,340
1964	32,180	4,718	8,454	45,352	6,571	698	3,789	397	374	285	12,104
1965	35,698	5,569	8,880	50,147	7,290	1,140	4,635	464	466	372	10,367
1966	40,359	5,040	10,683	56,082	8,075	1,011	5,316	536	512	492	15,942
1967	46,820	6,180	12,250	65,250	8,899	904	6,317	589	526	403	17,638
1968	55,507	5,902	14,094	75,503	10,486	1,318	6,781	685	569	662	20,501
1969	69,011	7,205	15,055	91,271	12,340	1,314	8,396	695	600	597	23,942
1970	78,164	5,909	17,275	101,348	12,686	1,342	9,141	675	591	721	25,156

Table 7. (cont.)

Year (1)	Graduates				S&T postgraduates						
	BSc (2)	BSc Agri. (3)	BE/BSc Engi. (4)	Total (2 to 4) (5)	MSc, incl. Home Science (6)	MSc Agri. (7)	MBBS (8)	MD (9)	MS (10)	ME/MSc Engi. (11)	Total (6 to 11)
1971	94,923	5,280	17,981	118,184	14,435	1355	9,285	778	666	807	27,326
1972	101,187	5,600	15,553	122,340	15,951	1496	9,524	895	525	720	29,111
1973	99,112	4,649	12,257	116,018	17,699	1469	10,239	1063	726	857	32,053
1974	108,287	4,505	11,430	124,222	17,437	1313	10,578	1161	799	818	32,006
1975	85,665	3,851	10,585	100,101	17,327	1500	10,098	1185	802	887	31,799
1976	83,577	4,700	10,912	99,189	16,408	1715	11,088	1323	867	790	32,191
1977	84,686	4,306	12,289	101,281	15,482	1719	11,963	1534	899	776	32,372

Source: University Grants Commission, *University Development in India*, part 2, 1983.

Table 8. India: Annual Ph.D. awards since 1950/1951

Year	Science	Agriculture	Engineering	Medical	Arts	Commerce	Total
1950/51	92	2	10	1	71	4	180
1951/52	109	5	12	–	82	3	211
1952/53	124	4	6	2	94	4	234
1953/54	164	4	19	1	111	10	309
1954/55	197	2	18	–	116	12	345
1955/56	217	11	24	9	146	4	411
1956/57	229	4	23	8	152	7	423
1957/58	241	8	20	16	213	6	504
1958/59	286	8	24	6	245	8	577
1959/60	351	11	18	11	286	26	703
1960/61	376	38	16	10	328	18	786
1961/62	401	41	20	14	314	15	805
1962/63	489	38	23	20	380	22	981
1963/64	516	52	26	17	420	15	1,046
1964/65	519	127	32	21	473	25	1,197
1965/66	683	92	39	39	541	20	1,394
1966/67	838	102	50	16	695	34	1,635
1967/68	984	89	66	20	769	31	1,959
1968/69	1,017	123	75	26	667	35	1,943
1969/70	1,103	157	98	24	788	46	2,216
1970/71	1,103	216	99	33	876	58	2,385
1971/72	1,168	182	76	46	929	47	2,448
1972/73	1,311	267	110	34	1,442	55	3,219
1973/74	1,327	276	95	46	1,093	58	2,895
1974/75	1,515	281	163	50	1,258	55	3,322
1975/76	1,516	289	136	42	1,282	41	3,306
1976/77	1,671	334	152	49	1,364	70	3,640
1977/78	1,901	329	168	51	1,674	96	4,219

Source: University Grants Commission, *University Development in India*, Part 2, 1983.

1,174,500 in 1970, 1,780,500 in 1980, and 2,406,700 in 1985, and is estimated to reach 3,022,300 in 1990. As a ratio, this was an increase from a mere 0.53 to 3.21 per thousand of population between 1950 and 1985.

By the end of the fifth Plan, S&T education, training and R&D infrastructure and manpower stock, and a range of industries had been set up in the country. The sixth Plan (1980–1985) recognized that "political independence has been matched by increasing technological independence in many areas." The Plan, however, also pointed to several deficiencies and mismatches, a lack of competitive capability,

Table 9. India's stock (economically active) of scientific and technical manpower, 1950–1990

	Economically active stock (in thousands)							
Manpower category (1)	1950 (2)	1955 (3)	1960 (4)	1965 (5)	1970 (6)	1980 (7)	1985 (8)	1990[a] (9)
Engineering and technology								
Degree (BE)	21.6	37.5	62.2	106.7	185.4	221.4 (254.5)[b]	324.2 (372.6)	395.3 (454.4)
Diploma	31.5	46.8	75.0	138.9	244.4	329.4 (378.6)	490.9 (564.2)	639.3 (734.8)
Science								
Postgraduates	16.0	28.0	47.7	85.7	139.2	217.5 (278.9)	273.2 (350.3)	327.4 (419.7)
Graduates	60.0	102.0	115.6	261.5	420.3	750.3 (961.9)	956.4 (1,226.1)	1,240.7 (1,590.6)
Agriculture and veterinary								
Postgraduates	1.0	2.0	3.7	7.7	13.5	96.5 (121.1)	128.6 (161.6)	156.1 (196.2)
Graduates	6.9	11.5	20.2	39.4	42.2	–	–	–
Medical								
Graduates (MBBS) and dental surgeons	18.0	29.0	41.6	60.6	97.8	165.4 (190.1)	233.4 (268.2)	273.5 (314.4)
Licentiates	33.0	35.0	34.0	31.0	77.0	NA	NA	NA
Total	188.0	292.7	450.0	731.5	1,174.5	1,780.5 (2,185.1)	2,406.7 (2,942.7)	3,022.3 (3,710.1)

Population (in thousands)	356,833	392,156	430,991	482,807	537,272	675,200	750,900	808,000
Stock of S&T manpower per thousand of population	0.53	0.75	1.04	1.52	2.19	2.64	3.21	3.74

Source: Cols. 2–6: CSIR Division of Scientific and Technical Personnel; Cols. 7–9: Planning Commission, *Sixth Five Year Plan, 1980–85*, p. 220, and *Seventh Five Year Plan, 1985–90*, p. 123; A. Rahman, *Science and Technology in India*, New Delhi: NISTADS, 1984. Population estimates from CSO, *Statistical Abstract, 1982*; 1985– estimate based on growth rate 1971–1981.

a. Estimated.
b. Figures in parentheses refer to total stock, economically active plus unemployed.

Table 10. Outlay for first to sixth plans (Rs. million)

	Plan outlay	Non-Plan outlay	Total outlay
First Plan	140.0	60.0	200.0
Second Plan	330.0	340.0	670.0
Third Plan	714.9	730.0	1,444.9
Annual Plan	471.5	834.4	1,305.9
Fourth Plan	1,422.7	2,313.2	3,735.9
Fifth Plan	10,332.9	5,349.3	15,682.2
Sixth Plan	19,194.1	14,477.8	33,671.9

insufficiency of S&T, and areas with no exploitative base. It called for "a detailed strategy for a major technological breakthrough appropriate to our resources and changing national environment."

The Plan also observed that the major investment areas in the plans required a much more deliberate and sustained application of S&T than hitherto. It called for "well-planned measures" in this regard so that technology "should strengthen the nation and reduce vulnerability."[11] The Plan emphasized the need to establish effective linkages in organizational forms and policy frameworks and the effective utilization of S&T to meet economic and social objectives.

The sixth Plan included several programmes, especially for "indicative thrust areas," and allocated Rs.19,194.1 million out of a total outlay of Rs.1,800 billion for the period 1980–1985. This was in addition to Rs.14,477.8 million non-Plan outlays, which brought the total to Rs.33,671.9 million. Though in percentage terms the public sector Plan outlay allocation (Rs.8,652 million) worked out at 0.9 per cent, the total amount allocated, given in normal terms, compared very favourably to allocation in the previous plans (table 10).[12]

Thus, the sixth Plan, whose outlay was about double that of the fifth Plan, aimed at technological breakthroughs.

During the sixth Plan period, the government issued the comprehensive Technology Policy Statement (1983), stating its objectives with regard to the development of indigenous technology and the efficient absorption and adaptation of imported technology appropriate to national priorities and resources. Its aims, *inter alia*, are making maximum use of indigenous resources, providing a maximum of gainful and satisfying employment, minimizing capital outlay, and promoting modernization, fuller capacity utilization, and energy efficiency. The Policy Statement noted the need for a system of efficient monitoring, review, and guidance, and a scheme of incentives and disincentives.

The seventh Plan (1985–1990) was formulated at a time when there was widespread concern about raising productivity in the country to meet the need for faster growth and a "resource crunch." The Plan emphasized the consolidation and modernization of the S&T infra-structure and the promotion of certain "thrust areas," and also the undertaking of work in frontline areas of new technology that had emerged. The Plan pointed to the need for S&T missions and linkages of S&T with the rest of the economy. Against the anticipated public sector outlay of about Rs.11,579 million in the sixth Plan (1980–1985), the seventh Plan envisaged an outlay of Rs.24,660 million, which works out at 1.4 per cent of aggregate public sector outlay for the Plan period 1985–1990.[13]

Table 11 shows S&T allocation in India's recent plans – the fifth (1974–1979), the sixth (1980–1985), and the seventh (1985–1990) – and reveals the national priorities in this sector, indicating the focus of the public sector on strategic and security areas. The direct share of the socio-economic sectors – industry, steel, energy, and agriculture – has been marginal. Atomic energy and space, which together accounted for 37.5 per cent of the allocations in the fifth Plan, had a smaller share in the sixth Plan (25.8 per cent), but more than recovered in the seventh Plan (41.2 per cent). The Council for Scientific and Industrial Research (CSIR) and the Department of S&T were allocated 18.4 per cent of the outlay in the fifth Plan, which fell to 15.9 per cent in the sixth Plan but rose to as much as 35.6 per cent in the seventh Plan. The seventh Plan envisaged increased linkages between S&T and industry and services.

Table 12, however, clearly shows that, compared to the projected growth in GDP in the Plan and the share of the socio-economic sectors in GDP, their direct allocations have remained disproportionately low, especially those for agriculture, manufacturing, and services.

Technology policy

The technology policy that has evolved over the last four decades has sought to protect local technology/skills from imported ones and to generate technology through direct and indirect policy instruments.[14]

Technology import

Import of disembodied technology
India's technology import policy is guided by the Prime Minister's Foreign Investment Policy Statement of April 1949, which recognized the role of foreign skills where local skills were not available. Foreign

Table 11. S&T outlay allocation in India's recent plans

Department/sector	Fifth Plan, 1974–79		Sixth Plan, 1980–85		Seventh Plan, 1985–90	
	Rs. million[a]	% of total	Rs. million[a]	% of total	Rs. million[a]	% of total
Atomic energy	1,671.3	21.77	2,489.8 (2,345.9)[b]	12.97	3,150.0	12.77
Space	1,282.7	16.71	2,458.0 (3,045.6)	12.81	7,000.0	28.39
S&T						
CSIR	817.7	10.65	1,700.0 (2,202.6)	8.86	3,350.0	13.58
D/ST	589.6	7.68	1,348.7 (2,699.3)	7.03	5,430.9	22.02
Subtotal	4,361.3	56.81	7,996.5 (10,293.4)	41.67	18,930.9	79.76
Socio-economic sectors						
Industry						
Heavy industry	287.6	3.75	575.1 (400.0)	3.00	NS	
Industrial development	103.2	1.34	177.0 (237.3)	0.92	NS	
Steel	66.2	0.86	417.0 (599.1)	2.17	NS	
Mines	64.8	0.84	161.6 (141.8)	0.84	302.4	1.23
Power	86.9	1.13	531.0 (284.5)	2.77	NS	
Coal	63.9	0.83	250.0 (61.5)	1.30	1,200.0	4.89

Electronics	187.3	2.44	323.4 (210.5)	1.68	NS
Communications	223.9	2.92	621.5 (405.7)	3.24	NS
Petroleum	120.8	1.57	390.8 (674.0)	2.04	1,917.4 (incl. petro-chemicals)	7.78
Chemicals and petro-chemicals	23.5	0.31	316.1 (NS)	1.65	NS
Agriculture ICAR	1,024.8	13.35	3,400.0 (2,871.0)	17.71	4,250.0	17.23
Others	69.1	0.90	120.0 (NS)	0.62	NS
Health and family planning	213.2	2.78	400.0 (480.9)	2.08	1,500.0 (ICMR)	6.08
Total (incl. others)	7,676.4	100.00	19,194.1	100.00	24,660.0	100.00

Source: Planning Commission, *Five Year Plans*.

a. At base year prices.
b. Parentheses show anticipated levels.
NS = Not specified.

33

Table 12. S&T outlay allocation and growth and composition of value added in the recent plans

Sector	Fifth Plan (1974–79)			Sixth Plan (1980–85)			Seventh Plan (1985–90)		
	R^a	S^b (1973–74)	STO^c	R^a	S^b (1979–80)	STO^c	R^a	S^b (1984–85)	STO^c
Agriculture	3.34	50.78	14.25	3.83	35.13	18.32	2.5	36.86	17.23
Coal	8.75	0.55	0.83	11.25 (mining)	NS	1.30	11.7 (mining)	NS	4.89
Petroleum	13.76	0.21	1.57						
Manufacturing	6.17	14.79	5.09	6.50	18.07	3.92	5.5	14.66	NS
Iron and steel	11.21	0.79	0.86	8.75 (basic metals)	1.26	2.17	5.5	2.04	NS
Electronics	7.57	0.05	2.44	NS	NS	1.68	NS		NS
Electricity	8.15	0.79	1.13	7.15 (including gas and water supply)	1.71	2.77	7.9	2.00	NS
Services	4.80	25.16	2.92 + 2.78 (community health and FP)	5.44	33.61	5.32	6.1	31.20	6.08 (health)

Source: Planning Commission, *Five Year Plans*.

a. R = Rate (%) of growth (annual compound) of GDP.
b. S = Share (%) in Gross Value Added.
c. STO = % allocation of total S&T outlay.
NS = not specified.

investment was considered a channel of technology transfer. However, it was considered desirable that Indians should have majority owner- ship and effective control, and the need for careful regulation in the national interest was emphasized. Vital importance was attached to the rapid indigenization and absorption of technical skills through the training of local technical personnel and the replacement of expatriates.

The foreign exchange crisis that the country faced in the late 1950s, coupled with a dearth of local entrepreneurs, led the government to adopt a more favourable attitude towards foreign collaboration and foreign investments. As the local fund of entrepreneurship, capital, and technology increased, the foreign collaboration policy was made more selective and restrictive in the late 1960s. The procedure of approvals of foreign collaboration proposals was streamlined in 1968.

A specialized body, the Foreign Investment Board (FIB), was created within the government to deal with all the cases involving for- eign investment or collaboration except those in which total invest- ment in share capital exceeded Rs.20 million and where the proportion of foreign equity exceeded 40 per cent; the latter were to be referred to the Cabinet Committee. A subcommittee of the FIB was em- powered to approve cases of foreign collaboration in which the pro- portion of foreign-held equity did not exceed 25 per cent and total equity investment Rs.10 million. The administrative ministries were authorized to approve cases involving purely technical collaboration.

In addition, the government, through three lists, separated areas (a) where no foreign collaboration was considered necessary, (b) where foreign technical collaboration was permissible, and (c) where even financial collaboration could be considered. Such lists were to be kept updated periodically. With regard to technical collaboration, maximum rates of royalty were specified for different items, with a maximum ceiling of 5 per cent, for a duration of normally five years. In order to ensure that foreign collaboration was avoided in areas where local technology might be available, local scientific agencies were represented on FIB and other screening bodies. Another guide- line issued by the government, also in 1968, was that wherever Indian consultancy was available, it was to be utilized exclusively. If foreign consultants were also required, Indian consultants were to be given the prime role.

The guidelines for foreign collaboration that evolved over time re- quired the importer to furnish the reasons for preferring the particular technology and its source. The technology imported should also be available for sublicensing within the country and have no minimum guaranteed royalty or restrictive clauses with respect to exports, source of capital goods, raw materials, or spares. Foreign brand-names should not be used for internal sales, and there should be a limit to renewals or extension of the collaboration.[15] In 1976, a Technical Evaluation Committee consisting of officials from the Council of Scientific and Industrial Research (CSIR), the Department of Science and Technology (DST), and the Directorate General of Technical Development (DGTD) was set up to assist the FIB in screening foreign collaboration proposals.

The government's decisions in 1970 regarding industrial policy, which were concretized in 1973, sought to restrict further the activities of foreign companies to a group of relatively complex technology and capital-intensive industries. The Foreign Exchange Regulation Act of 1973 imposed a general ceiling of 40 per cent on foreign shareholdings in Indian companies. Relaxations of this ceiling were given only to companies engaged in high technology or export-oriented activities.

The Technology Policy Statement (TPS) of 1983 placed emphasis on reducing technological dependence in key areas. Technology acquisition from abroad was not to be at the expense of the national interest, and due recognition and support was to be given to indigenous initiatives. The TPS contemplated the preparation and periodic updating of lists of technologies that had been adequately developed locally; normally no import of these would be permitted. The onus to demonstrate the necessity of that import was on the seeker. The TPS put a firm commitment on absorption, adaptation, and subsequent development of imported know-how through adequate investment in R&D, to which importers of technology were expected to contribute. A National Register on Foreign Collaboration (NRFC) was to be developed to provide analytical inputs at various stages of technological acquisition.

The initiatives taken as a follow-up to the TPS included a Technology Absorption and Adaptation Scheme (TAAS), which aimed at providing catalytic support for the accelerated absorption and adaptation of imported technologies. It was made mandatory for all importing firms to highlight steps taken towards the absorption of technology imports.[16]

Capital goods imports
Capital goods available locally are put on banned lists and cannot be imported. Some others, available on Open General Licences (OGL), can be imported by direct users. However, imports of more than Rs.1

million are permitted only after public advertisement, and when the authorities are satisfied that no local source exists.

In 1976, a Technical Development Fund was set up in the Ministry of Industry to promote the modernization of existing units. Under this scheme, an industrial unit can import US$500,000-worth of technical know-how, consultancy, design/drawings, or balancing equipment. In April 1985, the import duty on all project imports was slashed from 105 to 45 per cent; in the case of power equipment it was reduced to just 25 per cent, and was abolished on fertilizer equipment. Subsequently it was raised from 45 to 55 per cent.

Recent liberalization

Subsequent to the general trend of liberalization of the economy from 1979 onwards, the technology import policy has also been liberalized. The emphasis of the liberalization has been on modernization and the achievement of greater international competitiveness in order to increase manufactured exports. Capital goods imports policy has been liberalized by expanding OGL. Designs and drawings worth Rs. 200,000 could be imported against import replenishment (REP) licences without the government's prior approval. To encourage the modernization of exporting units, the Ministry of Commerce set up a Coordination Committee to sanction foreign exchange for the import of know-how, designs, and consultancy. The capital goods in respect of 13 specified core sector industries have been thrown open to global tender since 1978/79, irrespective of local availability. Policy guidelines were issued in November 1980 and subsequent months to streamline foreign collaboration approvals. Powers to approve technical collaboration agreements involving an outflow not exceeding Rs.500,000 were delegated to the administrative ministries.

Main features of technology import policy

Under the above technology import policies, a considerable volume of technology has entered the country over the last 38 years. Table 13 provides some indicators of technology imports for the period 1970/71 to 1985/86. It shows a significant increase in the volume of technology imported in the 1980s. The average number of foreign collaborations approved per year jumped from 270 for the period 1970/71 to 1979/80 to 660 for the period 1980/81 to 1985/86.

The technology import policy that evolved over the years has the following features. First, it is selective and seeks to provide protection

Table 13. Foreign collaborations approved and capital goods imports in India, 1970–1985

Year	Total no. of collaborations approved	Collaborations with foreign equity	Licensing/ technical assistance collaborations	Foreign investments approved[a]	Capital goods imports[a]	Of which non-electrical machinery[a]
1970/71	183	32	151	24.52	4,040	2,578
1971/72	245	46	199	58.38	NA	NA
1972/73	257	37	220	62.27	NA	NA
1973/74	265	34	231	28.17	NA	NA
1974/75	359	55	304	67.13	NA	NA
1975/76	271	40	231	32.05	9,677	5,767
1976/77	277	39	238	72.69	NA	NA
1977/78	267	27	240	40.03	11,484	6,947
1978/79	307	44	263	14.06	13,061	7,696
1979/80	267	32	232	56.87	14,585	8,069
1980/81	526	73	453	89.24	19,103	11,153
1981/82	389	57	332	108.71	20,961	13,492
1982/83	591	113	477	628.01	27,163	16,048
1983/84	673	129	547	618.70	29,814 (P)	19,738 (P)
1984/85	752	161	591	1,130.00	27,471 (P)	18,723 (P)
1985/86	1,024	238	786	1,219.00	NA	NA

Source: Indian Investment Centre; Government of India, *Economic Survey* (various years).

a. In Rs. millions.

to local technology where available. Local sources of all individual components of technology – consultancy, know-how, skills, and capital goods – receive protection from their foreign counterparts. Second, it seeks to reduce the direct and indirect costs of technology imports by regulating royalty and other payments. Third, it discourages packaged imports of technology. Technology import through foreign direct investments (foreign financial collaboration) is restricted only to select, relatively complex technology industries. Approvals of foreign financial collaboration are also subject to more stringent screening and attract ceilings on ownership. On the other hand, imports of designs/drawings/capital goods are less restricted. Finally, there is emphasis on rapid absorption, indigenization, and updating of the acquired technology.

Indigenous technological development

Parallel to the protection provided to the local technology, the government has taken several steps, in addition to its direct participation in scientific and technological development, to accelerate the pace of local technology generation. These steps include efforts to develop infrastructure for technological development, incentives to promote in-house R&D activity in industry, the enactment of a new Patent Act, and encouragement in the utilization of indigenous technology. The major initiatives taken in respect of each category are briefly discussed below.

Scientific and technological infrastructure

S&T MANPOWER TRAINING INSTITUTIONS. A large number of institutions training S&T personnel have been set up in the country, which now produces some 160,000 scientists and technologists per year (table 9).

POLICY BODIES. A number of policy-level bodies have also been set up to guide S&T programmes. The Scientific Advisory Committee to the Cabinet (SACC) was set up in 1956 and replaced by the Committee on Science and Technology (COST) in 1968. A National Committee on Science and Technology (NCST) was established in 1971 to formulate and continuously update comprehensive S&T plans. The first S&T plan formulated by NCST identified 24 major sectors for S&T development, and laid as much emphasis on the development of engineering, design, and fabrication skills as on the development of technology. The Science Advisory Committee to the Cabinet (SACC) was restored in 1981. In addition, there is a Cabinet Committee on Science and Technology. SACC has concerned itself with, among other things, the utilization of non-resident Indian technologists' support for S&T in the educational system, and the removal of regional disparities in scientific development.

NATIONAL LABORATORIES. A network of about 130 specialized laboratories and institutions, providing a variety of services such as testing, fabrication, consultancy, and measurements, have been established. Besides their own research, they operate within the framework of the research councils, i.e. the Indian Council of Agricultural Research (ICAR), the Council of Scientific and Industrial Research (CSIR), the Indian Council of Medical Research (ICMR), and other scientific departments such as the Department of Atomic Energy (DAE), the Department of Space, the Department of Electronics, the Department of Science and Technology (DST), the Department of Scientific and Industrial Research, the Department of Agricultural Research and Extension, and the Department of Environment, Defence Research, and Development Organization. These laboratories have proved to be the largest source of experienced scientists/technologists for in-house R&D activities in the industry.[17]

CONSULTANCY, ENGINEERING, AND DESIGN ORGANIZATIONS. The Indian government have taken initiatives to promote Consultancy, Engineering, and Design Organizations (CEDOs) in the country. In certain key technology-intensive sectors, CEDOs were promoted in the public sector, beginning in 1959 in metallurgy, with MECON (initially known as the Central Engineering and Design Bureau of the public sector Hindustan Steel Ltd.); in 1960 in fertilizers, with PDIL (initially the Planning and Development Division of the Fertilizer Corporation of India); in 1964 in general-purpose-machinery-utilizing sectors, with the National Industrial Development Corporation; and in 1965 in petroleum and related fields, with Engineers India Ltd. In addition, the government extended a variety of fiscal and other incentives to stimulate the growth of indigenous consultancy services. As a result, the country has about 300 firms offering a whole range of consultancy services, including project identification, techno-economic feasibility studies, market research, turnkey assignments, plant design and engineering, and project management, for almost any industry.[18]

Incentives to in-house R&D

Besides the creation of S&T infrastructure, the government has encouraged industries to take up in-house R&D activity through other policy instruments. In 1974, a scheme for the recognition of in-house R&D establishments of industrial units was started. The recognized R&D units get facilities for the import of equipment, raw material, samples, and prototypes under open general licence with no ceiling. The in-house R&D units are encouraged to work towards the following: import substitution, export promotion, process/product development and design improvement, development of new process/product technologies, and increased efficiency in the use of resources and fuels and the recycling of wastes.[19] Expenditure incurred on approved scientific research programmes is 100 per cent tax-deductible.

Sometimes foreign collaboration approvals are granted on the understanding that the importer will undertake R&D activity to absorb the technology. The New Drugs Policy (1978) obliged foreign companies with a turnover in excess of Rs.50 million to have R&D facilities within the country with capital investment of at least 20 per cent of their net block and to spend at least 4 per cent of their turnover on R&D. It also specified a 1–2 per cent higher profit ceiling for drug companies engaged in approved R&D work. In addition, the government encouraged and supported industries in setting up research associations to take on work on common problems. To promote technological diffusion, the industrial policy encouraged subcontracting by large units to small-scale ancillary units.

The new Patent Act

In the 1960s, foreign firms had used vague provisions of the patent law to hamper the innovative activity of Indian chemical and pharmaceutical firms.[20]

To avoid that possibility, a new Patent Act was enacted in 1970. The new Act abolished product patents in food, chemicals, and drugs and reduced the life of process patents from 16 to 7 years, and to 14 years in other cases. It contained provisions for a worldwide search of patent literature to establish the novelty of a product, or compulsory licensing after three years to preclude the situation in which foreign firms neither used their patents nor allowed them to be used.

The new Patent Act seems to have stimulated local innovative activity to a significant extent. The patent statistics provided in table 14 show that the number of patents filed and sealed in India has declined

Table 14. Number of patents from persons in India and abroad, 1957–1984/85

Year	Indians[a]		Foreigners[b]		Total	
	Applications made	Granted/ sealed	Applications made	Granted/ sealed	Applications made	Granted/ sealed
1957	609	–	2,847	–	3,456	–
1958	602	–	2,970	–	3,572	–
1959	726	–	3,239	–	3,965	–
1960	721	–	3,782	–	4,503	–
1961	774	–	4,515	–	5,289	–
1962	814	–	4,999	–	5,813	–
1963	878	–	4,798	–	5,676	–
1964	902	–	4,803	–	5,705	–
1965	948	–	5,054	–	6,002	–
1966	979	–	4,450	–	5,429	–
1967	1,125	–	4,065	–	5,190	–
1968	1,110	426	4,248	3,704	5,358	4,130
1969	1,120	645	4,326	4,308	5,446	4,953
1970	1,116	596	4,026	2,936	5,142	3,532
1971	1,231	629	3,114	3,294	4,345	3,923
1972	1,180	265	2,515	1,245	3,695	1,510
1972/73	1,143	278	2,496	1,064	3,639	1,342
1973/74	976	358	2,515	1,058	3,491	1,416
1974/75	1,148	737	2,258	3,207	3,406	3,944
1975/76	1,129	426	1,867	1,894	2,996	2,320
1976/77	1,342	928	1,762	1,964	3,104	2,892
1977/78	1,097	657	1,773	1,857	2,870	2,514
1978/79	1,124	281	1,808	1,000	2,932	1,281
1979/80	1,055	516	1,925	1,657	2,980	2,173
1980/81	1,159	349	1,795	670	2,954	1,019
1981/82	1,093	421	1,896	936	2,989	1,357
1982/83	1,135	405	1,950	822	3,085	1,227
1983/84	1,005	340	2,090	980	3,145	1,320
1984/85	1,001	263	2,318	1,206	3,419	1,469

Source: Controller-General of Patents, Designs, and Trade Marks, *Annual Reports*.

a. By Indian is meant a person or entity who/which has the status of a resident in India. Thus, non-Indians are equivalent to "non-residents."

b. Including foreigners resident in India.

since 1970. This is, however, not a reflection of a decline in innovative activity but only of a reduction in the proportion of frivolous applications that were made in the past to block competition. On the contrary, the Act has led to a significant and widespread upsurge in adaptive research and the copying of technologies, even by smaller firms, for the manufacture of various drugs, insecticides, herbicides, and other agro-chemicals. The reduced term of patents and free import of chemicals and drug intermediates has allowed many Indian companies to manufacture hundreds of essential products not possible before. It has led to a reduction in the time gap between the discovery of a drug elsewhere and its manufacture in India.[21]

Incentives for utilization of indigenous technology
The government has promoted the National Research Development Corportion (NRDC) with the specific responsibility of transferring technology from R&D laboratories to industry. NRDC commercializes the technologies developed with government support, undertakes further work towards upscaling laboratory know-how, setting up pilot plant, etc., and even provides risk finance for development projects. In addition, indigenous R&D utilization is promoted by various other incentives. Products based on indigenous R&D qualify for excise duty exemption.

The drugs and medicines developed indigenously do not fall within the purview of the Drugs Price Control order (1980) for the first five years. A higher rate (35 per cent) of investment allowance and depreciation is applicable to plant and machinery installed (between 1977 and 1987) for the manufacture of goods based on indigenous technology. Such products are exempt from the provisions of industrial licensing. Proposals based on indigenous technology enjoy a preferential treatment in industrial licensing. Royalties earned by Indian companies abroad through the export of indigenous technologies are completely free of tax, and those earned within the country are given a 40 per cent rebate. Furthermore, to inculcate technological entrepreneurship in the country, the government has recently proposed to launch a Venture Capital Fund (VCF) and is setting up a number of Science and Technology Parks (STPs). The VCF will be funded by a 5 per cent cess to be levied on technology payments abroad.[22]

R&D and self-reliance

As a result of the S&T policies, national expenditure on R&D and related activities increased rapidly. Total estimated R&D expenditure was Rs. 14,278.7 million in 1983/84 compared to Rs. 46.8 million in 1951/52 (table 15).[23] The major part of this expenditure was accounted for by the public sector, and as much as 80 per cent by the central government sector.

As a percentage of GNP, the R&D percentage has gone up from

Table 15. India: national expenditure on scientific R&D and related activities as percentage of GNP, 1950/51, 1955/56, 1965/66, and 1970/71 to 1983/84

Year (1)	Total R&D expenditure[a] (Rs millions)		GNP (Rs millions) at 1970/71 market prices (4)	R&D expenditure as % of GNP (3/4) (5)	Percentage share of central sector (6)	Population (millions) (7)	Per capita R&D Expenditure (8)
	Current prices (2)	1970/71 prices (3)					
1950/51	46.8	90.3	183,750	Neg.	100.00	359	0.25
1955/56	121.4	264.1	233,040	0.12	100.00	393	0.67
1960/61	–	–	270,540	–	–	434	–
1965/66	850.6	1,125.6	316,910	0.36	93.20	485	2.32
1970/71	1,680.2	1,680.2	399,790	0.42	85.84	541	3.11
1971/72	2,140.4	2,032.1	408,830	0.50	85.38	554	3.67
1972/73	2,340.5	1,997.5	405,900	0.49	76.93	567	3.55
1973/74	2,530.5	1,818.9	421,340	0.43	78.66	580	3.14
1974/75	3,240.3	1,978.4	423,150	0.47	79.91	593	3.34
1975/76	3,979.9	2,505.8	464,830	0.54	81.33	607	4.13
1976/77	3,909.6	2,304.4	470,400	0.49	79.39	620	3.72
1977/78	4,502.3	2,566.0	509,230	0.50	81.33	634	4.05
1978/79	5,586.9	3,116.6	543,470	0.57	79.39	649	4.80
1979/80	6,743.3	3,244.9	516,400	0.63	77.62	664	4.89
1980/81	8,136.4	3,513.1	551,750	0.58	75.86	679	6.79
1981/82	10,081.5	3,965.0	582,900	0.68	76.27	694	6.80
1982/83	12,375.6	4,566.1	605,180	0.75	77.86	709	6.44
1983/84	14,278.7[b]	4,817.4	650,730	0.74	77.60	724	6.65

Source: (1) Committee on Science and Technology, *Report on Science and Technology*; (2) Ministry of Information and Broadcasting, *India*, various annual issues; (3) CSO, *National Accounts Statistics, 1970/71 to 1983/84*, January 1984.

a. Including central, state, and private sector.
b. Estimated.

almost nothing in 1950/51 and 0.12 per cent in 1955/56 to 0.75 per cent in 1982/83 and 0.74 per cent in 1983/84. The percentage works out at 1.50 if we evaluate the R&D expenditure using the International Comparison Project conversion factors of the World Bank. Even before independence, eminent scientists had recommended allocating 1 per cent of national income per year to scientific research.[24] This goal has yet to be reached. Per capita real R&D expenditure increased from Rs. 0.25 in 1950/51 and 0.67 in 1955/56 to Rs. 6.45 in 1983/84, indicating a tenfold rise over 1955/56 and a twenty-six-fold rise over 1950/51.

In-house R&D units recognized by the Department of Scientific and Industrial Research (DSIR) numbered over 900 in 1983/84, including both the public and private sectors, compared to 11 at the time of independence, and 400 in the period 1975/76. They are estimated at present to have assets worth Rs. 6,000 million. These R&D units employed 45,000 people in 1983/84, compared to 13,000 in 1975/76. The main sectors covered by the units were chemicals, drugs, and pharmaceuticals (340), electrical and electronics (210), engineering industries (265), textiles (25), and others (84). As many as 226 units incurred expenditure of Rs. 2.5 million or more on R&D, including 71 that incurred expenditure of Rs. 10 million or more.[25]

In attempting to evaluate India's performance, we shall confine ourselves to the industrial sector. This is because in other socio-economic sectors, such as agriculture, atomic energy, space, or defence, there may be no alternative to local development or adaptation. In all these latter fields, India has achieved a considerable level of self-reliance. In agriculture, for instance, the successful application of new technology has enabled the country to achieve self-sufficiency in food.

There have been divergent views, from appreciative to highly critical, on the technological capability of India's industrial sector. This is because, first, a precise measurement of technological capability is not possible and value judgements are to some extent unavoidable. Second, there is much inter-industry variation in India's capabilities, depending upon factors such as the relative complexity of the technology involved, the rate of technological change, the market size, the rate of growth of industry in India, the market structure, government policy, S&T infrastructure, etc. Our purpose here, however, is to examine the trends with regard to the country's overall independence from foreign S&T inputs and the achievement of an autonomous S&T capability, which have been the principal objectives of S&T planning and policies.

Foreign technological dependence

There is no single precise indicator of foreign technological dependence, because technology is transferred under different forms, disembodied (such as foreign collaboration and transfers of designs and drawings) as well as embodied (capital goods and human skills). In a large country such as India, it is difficult to estimate the proportion of manufacturing output based on foreign or local technology. The number of foreign collaborations – purely technical ones as well as technical and financial – approved in a year is published by the government. But this can be a misleading indicator of technology import because a collaboration could be quite comprehensive, covering the entire product range, or it could be for a minor part or component. However, some idea of the extent of dependence on foreign resources by foreign collaborating firms can be gained from the relative significance of servicing payments remitted abroad. Table 16 summarizes the value of production and remittances and other earnings and expenditure in foreign exchange made by samples of companies having foreign collaborations in force from four successive Reserve Bank of India Surveys of Foreign Collaborations. It is apparent that total remittances in terms of dividends, royalty, technical fees, etc., by the foreign collaborating firms accounted for an average of 2.1 per cent of the value of production between 1960/61 and 1963/64. This proportion steadily came down to 1 per cent for foreign collaborations in force in the period 1977/78 to 1980/81. A part of this decline may be attributable to cost savings resulting from a possible improvement in India's bargaining capacity owing to the development of local technological capability. But in large measure it demonstrates that foreign collaborations have gradually become less broad-based and comprehensive. Alternatively, Indian firms over time may have become less dependent upon their foreign collaborators. This, in fact, is an important indicator of the capacity to unpackage technology imports.

Unpackaging of technology imports

Foreign direct investment (FDI) is considered to be a more packaged mode of technology transfer than licensing or purely technical collaboration agreements or imports of designs and drawings. Therefore, the relative importance of technology imports through FDI (or financial-cum-technical (F&T) collaborations) will reveal the trend towards unpackaging. Table 17 summarizes the proportion of F&T col-

Table 16. Remittances made by Indian companies in private sector having foreign collaboration

Year (no. of companies)	Value of production (A)	Remittances of dividend, royalty, etc. (direct cost) (B)	Value of import (C)	Total foreign exchange payments (total cost B + C) (D)	% share of remittances in value of production (E)	% share of total foreign exchange payments in value of production (F)
Period (827)						
1960/61	910.3	18.3	206.4	224.7	2.0	24.7
1961/62	927.1	22.2	212.8	235.0	2.4	25.3
1962/63	1,163.7	27.1	220.0	247.1	2.3	21.2
1963/64	1,420.1	26.6	242.8	269.4	1.9	18.9
Annual average	1,105.3	23.5	220.5	244.1	2.1	22.1
Period 2 (877)						
1964/65	1,570.3	30.6	216.0	246.6	1.9	15.7
1965/66	1,810.7	32.4	212.7	245.1	1.8	13.5
1966/67	2,112.6	42.5	286.0	428.5	2.0	20.3
1967/68	2,356.4	22.5	319.3	341.9	0.9	14.5
1968/69	2,559.3	52.2	279.8	331.9	2.0	12.9
1969/70	3,083.7	55.5	367.4	422.9	1.8	13.7
Annual average	2,265.5	39.2	280.2	336.1	1.7	14.8
Period 3 (859)						
1970/71	4,005.8	62.6	324.2	386.8	1.6	9.6
1971/72	4,529.7	58.6	399.8	458.4	1.3	10.1
1972/73	4,906.0	61.6	363.2	424.8	1.2	8.6
Annual average	4,480.5	60.9	362.4	423.3	1.3	9.4
Period 4 (594)						
1977/78	5,331.2	66.8	284.8	351.6	1.2	6.6
1978/79	6,230.9	70.1	372.6	442.7	1.1	7.1
1979/80	7,205.2	70.2	510.3	580.5	0.9	8.0
1980/81	8,739.5	72.5	624.9	697.4	0.8	7.9
Annual average	6,876.2	69.9	448.2	518.1	1.0	7.5

Source: Reserve Bank of India, *Foreign Collaboration in Indian Industry* (First, Second, Third and Fourth Survey Reports), Bombay: RBI, 1968, 1974, 1985, and 1985 respectively.

Table 17. Foreign collaboration approvals, 1961–1985

Year	Total number of cases approved	Of which financial and technical (F&T)	Proportion of F&T collaborations in total (% ages)
1961	403	165	40.94
1962	298	124	41.61
1963	298	115	38.59
1964	403	123	30.52
1965	241	71	29.46
1966	202	49	24.26
Annual average	307	108	35.18
1967	182	62	34.06
1968	131	30	22.90
1969	134	29	21.64
1970	183	32	17.45
1971	245	46	18.77
1972	257	37	14.40
1973	265	34	12.83
1974	359	55	15.32
1975	271	40	14.76
1976	277	39	14.08
1977	267	27	10.11
1978	307	44	14.33
1979	267	32	11.98
Annual average	242	39	16.11
1980	526	73	13.88
1981	389	57	14.65
1982	592	113	19.09
1983	673	129	19.17
1984	752	151	20.08
1985	1,024	239	23.34
Annual average	659	127	19.27

Source: Department of Scientific and Industrial Research.

laborations approved by the government of India between 1948 and 1985. It is apparent that the proportion of such collaborations has steadily fallen from an average of 6.36 per cent during the period 1961–1966 to 16.11 per cent for 1967–1979. The liberalization of policy during the 1980s has slightly increased their share to 19.27 per cent over the past five years. Part of this decline may be illusory, as in the early 1960s F&T collaborations were also favoured because of foreign exchange and capital scarcity. Nevertheless, it also shows that over the years Indian industry has acquired the capability gradually to unpackage technology imports.

Absorptive capacity

A large component of technology is tacit, residing in the skills of technicians who implement the projects and operationalize them. The indigenization of technicians will, therefore, be an indicator of absorption of "know-how," if not of "know-why." Table 18 shows the number of foreign technicians employed by companies engaged in foreign collaboration, again compiled from the successive RBI surveys. It is apparent from the table that there was on average one foreign technician per Rs.6.1 million of value added during the period 1964/65 to 1969/70. The average value added per foreign technician amounted to Rs.168.5 million during the period 1977/78 to 1980/81, representing an increase of over 26 times. A part of this increase is, however, due to inflation, as value added is measured in current prices. The significance of total payments to foreign technicians in value added (both measured at current prices) is a better indicator of the extent of dependence on them. This proportion has fallen steadily from an average of 0.95 per cent during the second survey period to 0.05 in the latest survey – a trend that suggests a considerable decline in dependence on foreign manpower. The table also shows a decline in the number of Indian technicians sent abroad for training as a proportion of value added under foreign collaborations over the years.

Local consulting, design, engineering, and equipment fabrication capability

Over the past three decades the country has built up an engineering industry that can manufacture a very wide range of equipment in virtually every industry. The value added in capital goods industries increased by three-and-a-half times in real terms between 1960 and 1980. As a result of the development of the local capability to design, engineer, and fabricate equipment and of local consultancy services, the country's dependence on imports of capital goods has fallen drastically. Table 19 shows that imported machinery and equipment accounted for 61.7 per cent of gross domestic capital formation in machinery and equipment in 1960/61; the proportion at constant prices had come down to a mere 6.7 per cent by 1982/83.

Table **18.** Foreign technicians employed and Indian personnel sent abroad for training by Indian companies having foreign collaboration

Year (no. of companies)	Value added (Rs. crores)	No. of foreign technicians	No. of trainees sent	Total payments to foreign technicians	Value added per foreign technician	Value added per trainee sent abroad	% share of payments to foreign technicians in value added
Second survey (877)							
1964/65	517.9	1,309	506	6.04	0.40	1.02	1.17
1965/66	579.1	1,339	545	6.69	0.43	1.06	1.15
1966/67	651.1	1,404	590	7.53	0.46	1.10	1.16
1967/68	741.8	1,165	533	7.62	0.64	1.39	1.03
1968/69	852.8	980	549	7.04	0.87	1.55	0.82
1969/70	988.6	884	650	6.34	1.12	1.52	0.64
Annual average	721.9	1,180	3,373	41.26	0.61	1.28	0.95
Third survey (859)							
1970/71	1,336.1	708	584	7.08	1.89	2.29	0.53
1971/72	1,496.6	641	599	5.02	2.33	2.50	0.33
1972/73	1,687.0	560	569	4.43	3.01	2.96	0.26
Annual average	1,489.9	636	1,752	16.53	2.34	2.55	0.36
Fourth survey (594)							
1977/78	1,829.0	115	392	0.95	15.90	4.66	0.05
1978/79	2,100.0	134	471	1.02	15.67	4.46	0.05
1979/80	2,394.4	132	496	1.31	18.14	4.83	0.05
1980/81	2,779.7	159	515	1.58	17.48	5.40	0.06
Annual average	2,275.8	135	1,874	4.86	16.85	4.86	0.05

Source: Reserve Bank of India, *Foreign Collaboration in Indian Industry* (First, Second, Third and Fourth Survey Reports), Bombay: RBI, 1968, 1974, 1985, and 1985 respectively.

Table 19. Share of imports in India's fixed capital formation in machinery and equipment, 1960/61 to 1982/83 (Rs. millions at 1970/71 prices)

Year (1)	Gross domestic fixed capital formation in machinery and equipment (2)	Imports of machinery and transport equipment (excluding metal manufac.) (3)[a]	Index of (3), 1970/71 = 100 (4)	Percentage share of (3) in (2) (5)
1960/61	15,130	9,330	236	61.7
1961/62	16,720	9,470	240	56.6
1962/63	19,340	9,470	240	49.0
1963/64	21,390	9,090	230	42.5
1964/65	24,750	10,810	274	44.0
1965/66	25,530	9,640	244	37.8
1966/67	21,830	7,070	179	32.4
1967/68	21,530	8,070	204	37.5
1968/69	22,300	7,310	185	32.8
1969/70	23,700	5,340	135	22.5
1970/71	23,460	3,950	100	16.8
1971/72	27,110	4,900	124	18.1
1972/73	29,870	4,710	119	15.8
1973/74	33,570	5,340	135	15.9
1974/75	32,860	4,120	104	12.5
1975/76	33,690	4,190	106	12.4
1976/77	38,350	4,240	107	11.0
1977/78	41,050	5,070	128	12.4
1978/79	44,290	4,480	113	10.1
1979/80	45,610	4,070	103	8.9
1980/81	49,020	6,110	155	12.5
1981/82	53,820	5,170	131	9.6
1982/83	61,620	4,140	105	6.7

Source: CSO, *National Accounts Statistics*, various issues, and DGCI&S, Ministry of Commerce.

a. Deflators for column 3 based on 1968/69 base index until 1980/81 and on 1978/79 base index for the following years.

A World Bank team noted that the Indian non-electrical industrial machinery industry can supply complete, economic-sized units to the cement, sugar, and thermal power industries; about 80 per cent of the machinery requirements for large-sized paper and pulp plants; and 50 to 60 per cent of the machinery to chemical industries. The Indian machinery manufacturing plants were rated to be in very good condition and compared favourably with their Western counterparts in the use of labour and other inputs. The team was impressed with the "high

51

quality and professional capability of management in all plants, whether owned by the private or the public sector." It also found the "status of manufacturing technology adequate for the type of product manufactured in each case, though the level of sophistication varied substantially among plants." Several leading plants "are technologically as competent as similar plants in the industrialized countries" and have been pre-qualified by international consulting engineers and contractors for the manufacture of mechanical equipment and participation in international competitive bidding. The leading manufacturers "produced equipment and machinery of competitive international quality and were up to the standard of world equipment producers in manufacturing capacities and in efficiency of raw material use."[26]

To evaluate the economic efficiency of the Indian machinery industry, the team calculated the effective protection coefficient and domestic resource costs of local production. From these calculations, India emerges as an efficient manufacturer in all but three of the 19 categories of equipment studied. Despite higher domestic costs of inputs, the output prices for many items were found to be significantly lower in India than abroad. The situation would have been better had there not been net disincentives to the sector because of greater protection on inputs than on outputs.[27]

A more recent study of Indian power and fertilizer equipment manufacturers has also revealed that, had the Indian manufacturers obtained inputs (steel, non-ferrous metals, alloys, power) and capital at the same price as their foreign counterparts, their prices would compare well with those of imported ones in the case of fertilizer equipment and would even be lower in the case of power equipment.[28] Therefore, the allegedly high prices of Indian-made capital goods, if they are indeed high, are due not so much to technological or economic inefficiency as to high input prices. In fact, it is on the strength of the well-diversified and competent engineering goods industry that India emerged as a significant exporter.

It is apparent from table 20 that exports of engineering goods rose from a negligible Rs.52 million in 1956/57 to Rs. 9,389 million in 1981/82. Subsequently the growth of exports of engineering and capital goods stagnated owing to the global recession, in particular in the Middle East countries. In some specific capital goods such as boilers (power), India accounted for about 2 per cent of world exports and was well ahead of other developing countries such as Brazil, Colombia, Argentina, the Republic of Korea, and even Spain and Yugoslavia.[29] In addition, Indian engineering firms have been exporting an increas-

Table 20. India's technology-linked exports, 1956/57, 1970/71, and 1973/74 to 1984/85 (Rs. millions at current prices)

Year (1)	Total exports (2)	Engineering goods (3)	Capital goods[a] (4)	Electronics[b] (5)	Consultancy services (6)	Computer software (7)	% of (3) in (2) (8)
				Technology-linked exports			
1956/57	6,045	52	62	Nil	Nil	Nil	0.9
1970/71	15,352	1,304	–	11 (1971/72)	10	–	8.5
1973/74	25,234	2,017	639	93	13	–	8.0
1974/75	33,288	3,566 (3,491)[c]	1,116	116	15	–	10.7
1975/76	40,428	4,009 (4,082)	1,552	162	39	–	9.9
1976/77	51,423	5,663 (5,517)	1,773	220 (271[d])	75	–	11.0
1977/78	54,079	6,174 (6,240)	2,025	369[d]	95	–	11.4
1978/79	57,261	6,044 (7,169)	2,549	396[d]	140	–	10.6
1979/80	64,184	7,391 (7,367)	2,789	466[d]	220	–	11.5
1980/81	67,107	8,150 (8,742)	3,044	418[d]	251	72[d]	12.0
1981/82	78,059 (76,099)[e]	9,389 (10,470)	3,664	564[d]	430	102[d]	12.0 (12.3)
1982/83	88,053 (77,400)	7,992 (10,113)	5,150	890[d]	518	135[d]	9.0 (10.0)
1983/84	98,721 (86,410)	6,914 (11,700)	4,215	1,070[d]	630	240[d]	7.3 (8.0)
1984/85	116,569 (100,938)	7,384 (11,500)	4,300	1,170[d]	600	289	6.3 (7.3)

Source: Columns 2 and 3: DGCIS, Ministry of Commerce, and Ministry of Commerce, Annual Reports; columns 4–7: Engineering Export Promotion Council (EEPC), Association of Indian Engineering Industry (AIEI), and Department of Electronics (Annual Reports).

a. Covering machinery (electrical and non-electrical) and transport equipment.
b. Including those of Santacruz Electronics Export Processing Zone (SEEPZ), part of column 4.
c. Parentheses in column 3 show estimates of EEPC/AIEI.
d. Refers to calendar year.

ing volume of consultancy services. Starting from exports worth a mere Rs.10 million in 1970/71, the figure rose to Rs.630 million in 1983/84.

India, therefore, has acquired substantial consulting, design, engineering, and equipment fabrication capabilities, which have enabled it not only to become self-reliant to a considerable extent in the field of S&T, but also to export a significant volume of equipment and technology to co-developing countries.

Adaptation and indigenization capability

The technologies imported may not be appropriate to local factor proportions and available raw materials. Hence they have to be suitably adapted and indigenized to maximize their benefit to the nation. Furthermore, the process of adaptation is potentially an important source of knowledge and absorption of "know-why." There is evidence that considerable adaptation of the technologies being imported by Indian firms has been undertaken, for various reasons: to facilitate the substitution of imported raw materials by local ones; to adapt the product to the local climate, environment, or tastes; to scale down plant size to suit the size of the market; to upgrade the technology, etc. These adaptations have resulted in indigenization of the imported technologies.

An indicator of the indigenization of technologies is the evidence of technology exports from India to other developing countries through joint ventures and licensing agreements.[30] A survey of 52 Indian firms who undertook joint ventures abroad revealed that though 42 of them had obtained their initial technology from foreign sources, 47 of them had indigenized them beforehand.[31] That locally adapted technology was exported to their affiliates is apparent from the fact that "Indian machinery" was the source of know-how in most of the Indian investments.[32] The adaptations made in Indian technologies make them "most appropriate" for conditions in developing countries.[33] An empirical study for 12 four-digit Thai industries confirmed that Indian firms "often used absolutely efficient technologies; they operated to the left of the production isoquent with costs below the theoretical minimum cost."[34]

Product development and innovations

The capacity to innovate is the highest stage in the accumulation of knowledge of technology by a country. The Indian S&T infrastruc-

Table 21. Patents filed and sealed in India, 1968–1985

Year (1)	Indians Applications made (2)	Sealed (3)	Foreigners[a] Applications made (4)	Sealed (5)	Percentage share of Indians in total patents sealed (6)
1968	1,110	426	4,248	3,704	10.31
1969	1,120	645	4,326	4,308	13.02
1970	1,116	596	4,026	2,936	16.87
1971	1,231	629	3,114	3,294	16.03
1972	1,180	265	2,515	1,245	17.55
1972/73	1,143	278	2,496	1,064	20.07
1973/74	976	358	2,515	1,058	25.28
1974/75	1,148	737	2,258	3,207	18.69
1975/76	1,129	426	1,867	1,894	18.36
1976/77	1,342	928	1,762	1,964	32.09
1977/78	1,097	657	1,773	1,857	26.13
1978/79	1,124	281	1,808	1,000	21.93
1979/80	1,055	516	1,925	1,657	23.74
1980/81	1,159	349	1,795	670	34.25
1981/82	1,093	421	1,896	936	31.02
1982/83	1,135	405	1,950	822	33.01
1983/84	1,005	340	2,090	980	25.76
1984/85	1,001	263	2,318	1,206	17.90

Source: Controller General of Patents, Designs, and Trade Marks, *Annual Report.*
a. Including foreigners resident in India.

ture – i.e. in house R&D units and national laboratories – as well as adapting and indigenizing foreign technologies, has started innovating. In some areas, these efforts have yielded tangible outcomes.

For instance, of the 34 models of tractors sold in the country, as many as 16 are products of in-house R&D, and one was developed by a national laboratory.[35] Other significant product developments include battery-operated electric locomotives, electronic control equipment for diesel engines, and high-speed power thyristors; solar photovoltaic cells and solar-powered electronic controls for offshore oil-exploration vessels; instrumentation systems for improving the productivity of sugar mills; catalysts for the petroleum-refining and fertilizer industries; a whole range of modern pesticides with low residual effects; light compact aircraft[36]; low-cost digital switching technology for telecommunications; and the technology for electronic-grade silicon. In the last two cases, the locally developed technology has proved to involve a much smaller capital outlay than the available imported alternatives, hence resulting in lower product costs.

An indicator of increasing innovative and product development capability is the proportion of patents held by nationals. Table 21 gives Indians' share of total patents for the period 1968 to 1984/85. Though the proportion fluctuates from year to year, the trend suggests that it has gone up significantly, from about 15 per cent during the early period (up to the early 1970s) to about 30 per cent in the 1980s. This doubling of resident Indians' share of patents suggests a significant improvement in India's capability to innovate and develop new products and processes. A part of the apparent increase may, however, be attributable to the amendment of Indian patent legislation, which has reduced the life of patents. It is possible that after the revision foreign firms find it less attractive to obtain a patent in India.

India's technological capability: An international comparison

Intercountry comparisons of technological capability are fraught with difficulties because of lack of a precise indicator. One broad classification divides countries into three basic categories:
– Technology leaders: The United States, Japan.
– Technology followers: Other OECD countries.
– Technology borrowers: Developing countries.[37]
More detailed classifications are possible either on the basis of indicators of S&T potential or on those of performance.

A study by Konrad and Wahl classified countries into eight stages of S&T development on the basis of seven quantitative indicators of S&T potential, such as S&T personnel, R&D expenditure, proportion of productive R&D, etc., and five qualitative indicators of development of S&T infrastructure (table 22).[38] According to these criteria, India is classified in the sixth stage (of the eight successive stages), along with other newly industrializing countries (NICs) such as Brazil, the Republic of Korea, Indonesia, and Argentina. The higher stages, i.e. seventh or eighth, include all developed countries, with the USA, Germany, Japan, and France in the eighth, and Italy, the Netherlands, and Canada in the seventh.[39]

Another study analysing the technology exports of developing countries has rated India as "most diverse and 'deep' (in terms of going into basic design of products and capabilities among the NICs . . . India is able to provide not just the operating knowledge to set up and run industries (the know-how), but also the design and manufacture of the plant and equipment, designed specifically for the client (the know-why)."[40] India's achievement is considered to be more notable because

56

Table 22. Stages in scientific and technological development

Stage	Infrastructure science and management	Higher educational institutions	Research and development	Services	Societies	No of researchers	Researchers per million inhabitants	R&D expenditure	R&D expenditure per capita	Share of R&D expenditure in GDP (%)	Researchers in productive sector (%)	R&D expenditure in prod. sector (%)	Countries (examples)
8	V	V	V	V	V	73,000–661,000	1,365–3,940	10,000–70,000	195–305	1.8–2.5	56–71	61–71	USA, FRG, Japan, France
7	V	V	IV	IV	IV	26,000–49,000	824–1,869	2,450–3,500	25–450	0.2–1.3	44–60	47–66	Italy, Netherlands, Canada
6	IV	IV	IV	III	III	9,500–28,000	89–535	430–1,160	1.0–27.0	0.5–0.7	13–35	31–66	India, Brazil, Republic of Korea, Indonesia, Argentina
5	III	III	III	III	II	1,600–5,600	61–188	30–205	0.5–16.5	0.2–0.6	16–68	20–82	Pakistan, Philippines, Venezuela, Ecuador
4	II	III	II	II	II	1,500–4,000	31–247	13–158	0.8–3.7	0.3–	~25	~45	Nigeria, Peru, Ghana, Sudan, Burma
3	I	II	I	II	I	94–1,500	27–537	0.3–22.0	0.2–19.5	0.2–0.4	3.5–70.0	6.8–75.0	Guyana, Mauritius, Kenya, Japan, Kuwait
2	O	I	O	I	O	75–500	13–100	0.4–18.8	0.5–3.7	0.1	0	0	Niger, Lebanon, Central African Rep., Togo, Rwanda
1	O	O	O	I	O	2–10	31–50	0.1–2.0	–	–	–	–	Seychelles, Benin, Gambia

Source: Norbert Konrad and Dietrich Wahl, "Scientific and Technological Potentials in Developing Countries as National and International Economic Factors," *Economic Quarterly* 20 (1985); based on UNESCO, *Statistics on Science and Technology*, Paris, 1983.

V = highly integrated network, complex spectrum; IV = network, close contacts of elements, limited spectrum; III = developed institutions in key areas with contacts; II = several institutions with contacts; I = various institutions; O = no potential.

of the fact that "India seems to have the lowest relative reliance on foreign technology of all the NICs in the past fifteen years or so."[41]

The inference that can be gained from the two studies cited above is that India, though far behind the developed countries, has one of the most advanced S&T capabilities among the developing countries.

Policy perspectives

Although India has made considerable progress towards the goal of building up an autonomous S&T capability, its achievements have still fallen short of expectations or potential. Thus the proportion of projects, particularly in large and modern industries, based entirely on indigenous technology continues to be small. In spite of the growing amounts spent, the national laboratories have failed to become important sources of industrial technology.[42] Furthermore, a large proportion of the technologies developed by them are not utilized for commercial purposes.[43] Despite a number of fiscal incentives granted by the government, only about 900 recognized in-house R&D units existed in 1986. Those spending Rs.10 million or more per year numbered only 71.[44]

The technology policy, therefore, in spite of considerable planning and effort, seems to have had only limited success in promoting industrial R&D and the utilization of local technology. Taking these considerations into account, we outlined below some proposals with regard to future technology policy, to make research in national laboratories more effective, to promote in-house R&D, and to facilitate the utilization of local technology.[45]

Research in national laboratories

The impulse to innovate originates either in the production process (process simplifications/improvements, material/labour-saving devices, overall efficiency, etc.) or in the market (product improvement, changing demand/tastes/patterns, scope of import substitution, etc.). The national laboratories receive no feedback from either of these sources. They operate in an environment isolated from production units, do not sell innovations directly to the industry, draw their resources almost entirely from the government, and have mostly officials and scientists on their boards. They are not involved in the import of technology. Therefore the possibility of their contributing to the local absorption of imported technology does not arise.

The need for more intensive links with industry has been emphasized repeatedly, but such links have largely been limited to occasionally accepting industry-sponsored research. There are two possible ways of making their research more fruitful. First, they could work more closely with the public sector enterprises, as recommended by the Working Group on Organization of R&D within the Public Sector and Relationship with the National Laboratories.[46] Public sector enterprises account for nearly half the capital investment in the country and operate across a wide range of relatively high-technology, capital-intensive and process-based industries. Except for a few leading ones, such as BHEL, HMT, BEL, and consultancy and design organizations, their record in R&D is generally poor. National laboratories could help these public sector enterprises to assimilate and adapt imported know-how and to generate their own. Public ownership of both enterprises and laboratories should make it administratively easier to implement this proposal.

A second possibility is for national laboratories to be made to survive more and more on sponsored research from industry. This might exclude projects with high social externalities, such as those concerning rural industrialization, basic needs, etc. In addition, national laboratories could be involved in all major deals of selection, transfer, and adaptation of technology.

In-house R&D

Empirical studies of in-house R&D in industry have identified its two important determinants: (1) market structure and (2) mode of technology imports besides structural factors, such as technological opportunities in the sector and product market characteristics. In addition, a firm's decision to "make" (or develop) technology locally or "buy" (import) it from abroad would naturally be based upon the relative costs of these two options.

Market structure

In-house R&D activity has been related to market structures in the framework of the Schumpeter's theory of creative destruction.[47] The basic proposition of the Schumpeterian and neo-Schumpeterian paradigms is that oligopolistic market structures are more conducive to R&D than pure or perfectly competitive ones. R&D, along with advertising, is taken to be an entry barrier, raising the investment undertaken by the leading firms to protect their market shares. But if the threat of further entry is absent, then oligopolistic market structures may not lead to R&D activity. This is precisely what seems to have happened in a number of Indian industries.[48] The early industrial licensing policy did lead to oligopolistic market structures, but also eliminated the threat of further entry. With trade policies ruling out any import competition, the "policy entry barriers" which protected the industry's firms were almost invincible, except in very high growth industries. Hence firms did not feel any need to carry out R&D. The technology policy, therefore, ought to take note of appropriate market structures. It must be pointed out, however, that the market structure most appropriate for spurring technological change may be different for different industries, depending on factors such as economies of scale in production and innovation.[49]

Technology import
The nature of the relationship between technology imports and local in-house R&D in the Indian context has been found to be dependent upon the mode of import, i.e. whether by licensing agreement or foreign direct investment (FDI). A number of empirical studies, at both firm and industry level, have confirmed that firms importing technology through FDI are much less concerned with absorption, adaptation, and in-house R&D than their counterparts with technology imported under licensing agreements.[50] Hence, from the point of view of promoting indigenous technological capability through faster absorption and innovation, the policy ought to restrict technology imports through FDI. The desirability of the current trend of liberalization of foreign collaboration approvals in favour of financial ones, therefore, needs to be reassessed from this point of view.

Cost-effectiveness of local generation
A rational firm's decision to "make" or to "buy" a technology abroad is expected to be based upon the cost-effectiveness of local generation. The cost of imported technology for the importer is likely to be lower than the cost of local development. This is because the transfer of already developed technology, which is a public good, does not entail many costs, while fresh generation certainly does. However, there are externalities involved in local generation: the benefits latent in generation, such as skill formation, instant absorption, overall technological capability-building, are all available to the country. With technology imports, on the other hand, there are certain costs to the society other than the direct cost. In other words, the market price of imported

technology underplays its real cost, while that of local technology overstates its real cost.

The existence of these externalities in the technology market calls for state intervention. The state intervention that exists in the area of technology import is the entry regulation, which seeks to limit foreign collaboration in the areas where local alternatives are available. This "stop" and "go" protection accorded to local technology is deficient in many respects. For those technologies which manage to set themselves on the "stop" list, an almost total lack of foreign competition, or the threat of it, may take away any incentive to keep up to date. Hence, obsolescence may creep in. In the case of technologies that are on the "go" list, local generation would never be cost-effective, as we observed above. Hence, there would be no incentive to generate them locally. Thus, present policy does not protect the potential technology and hence retards innovative activity. It is, therefore, desirable to provide price protection to local technology, resembling the tariffs imposed on the import of goods. Operationally, it would mean imposing a tax or duty on the importers for any payments made to import technology. For instance, for every Rs.100,000 remitted abroad as a lump-sum fee, the technology importer would have to pay another Rs.100,000 (if it is 100 per cent protection) to the Exchequer. This would increase the effective cost of imported technology to the importers *vis-à-vis* the local technology, and might thus offset part of the price disadvantage the local technology faces without affecting the earnings of the technology supplier.

The Long-term Fiscal Policy (1985) has proposed a 5 per cent levy on payments for technology imports to augment resources for the Venture Capital Fund (VCF). But the objective of the proposal seems to be merely to raise resources for the VCF rather than to provide price protection to local technology, since a rate of 5 per cent would make only a marginal difference to the cost of imported technology. Like tariffs on goods, different rates of price protection may be set for different classes of technologies depending upon the potential of local development and the intensity of the need.

Utilization of local technology

Like the generation, of indigenous technology, its utilization also suffers from several disadvantages *vis-à-vis* imported technology on account of the following factors.

61

Time, capital cost and uncertainty

The "productionizing" of a standardized imported technology by experienced personnel may require a considerably shorter time than the commercialization of an indigenously developed technology from scratch. The former is also subject to less uncertainty and risk of failure because it has been proved and standardized. Furthermore, with imported technology it is possible to phase the project cost over a period of time. Normally, such projects begin with the assembly of imported kits, and the manufacturing process is indigenized gradually as markets are developed with the products assembled from the kits.

In contrast, an entrepreneur using indigenous technology has no such option. He must provide for the entire project cost at one go and develop markets from scratch. Once he enters the project, he cannot quit if the market does not pick up, while the one still selling the product assembled from imported kits can. The choice of local technology in preference to foreign alternatives, therefore, may prove to be time-consuming, more capital-requiring and subject to greater risk. Public policy should devise some instrument to offset these disadvantages to make utilization of local technology more attractive.

Finance – technology nexus

The ability of technology and equipment suppliers to provide financing (suppliers' credits) plays an important role in technology selection, particularly for large capital-requiring projects. Technology suppliers from industrialized countries are usually willing to provide or arrange financing on soft terms from their respective country's export-import banks or other institutions. Their bids are often backed up by their home government's bilateral aid agreements. The local technology or equipment suppliers with no matching ability to provide credits are, therefore, easily outmatched, even with comparable prices and capability. In the past few years, local engineering industry has, indeed, lost numerous orders to foreign firms in fertilizers, power, and steel projects because of lack of finance. It is, therefore, imperative that a fund be created to provide financing for projects using local technology and equipment to mitigate the problems of local suppliers in providing credits.

Market power of foreign technology

The prospect of using an internationally reputed brand- or trade-name gives a tremendous edge to foreign technology over the local ones, particularly in consumer goods. Though the guidelines for foreign collaboration stipulate that no foreign brand-names will be allowed to be used in domestic sales, they are very much in use. In fact, a number of foreign collaborations are just "cover-ups" for the procurement of the right to use foreign brand-names, and are being signed even in low-priority industries such as cigarettes. In order to make sure that only genuine technology is transferred to the country and the local technology does not face unfair competition from the market power of foreign technology, foreign brand-names have to be eliminated altogether. The present Trade Marks Act is seemingly vague on what constitutes a foreign brand and hence needs amending.

A useful way to define this would be to consider as foreign any brand-name that was in use abroad before its registration in India and any owned by foreign organizations, whether or not any royalty is paid for its use.

Case-studies

The distribution of technological capabilities across different industries is uneven and depends upon several policy and structural factors. As a part of this study, detailed case-studies were made in the machine-tools, power-equipment, petroleum-refining, and fertilizer industries. These detailed industry studies assessed the level of S&T self-reliance achieved, and also the role played by different factors, especially government policy. The conclusions of these studies are summarized below.

Engineering industries

In engineering industries, broadly three successive stages in the development of S&T capability can be delineated: (a) the capability to manufacture products in accordance with the specifications laid down by foreign manufacturers; (b) the capability to absorb, adapt, and modify foreign designs to specific conditions in the factor and product markets in developing countries; and (c) the ability to understand and/or develop basic principles of technology and to innovate.

Machine tools
A study of technological self-reliance in the machine-tools industry is of interest because it is an industry where engineering design rather

than science-based innovation plays a crucial role in technological development. Besides, this industry has had a long history of growth in India, beginning as early as the 1940s.

In the early 1950s, the decision to protect the Indian machine-tools industry from competing imports gave an initial impetus to growth based on import substitution. The decision to set up Hindustan Machine Tools (HMT) at that time provided a nucleus for the generation of technological capabilities and skills. With regard to the prevailing industrial environment and the available skills and technologies, the creation of HMT not only provided the impetus for the machine-tools industry but also foreshadowed the development of the engineering industry in the country.

By the mid-1960s, this industry had acquired the capability to absorb imported technology and to manufacture machine tools to the specifications laid down by foreign collaborators. Under the pressure of the recession of the mid-1960s, the industry undertook the more complex tasks of modifying machine tools and developing variants of machines for which the design had been acquired by the purchase of licences.

The drastic import-control regime during the 1950s, 1960s and, to some extent, the 1970s led to high growth rates in production, but did not necessarily bring about adequate technological advances, and a technology gap developed. The cost-competitiveness of the industry also suffered a set-back owing to its lack of exposure to international competitive forces.

More recent changes in the tariff policy, import-control policy, and technology-import policy of the government has led to a correction of these distortions. The competing imports of machine tools were increasingly allowed on quality and cost considerations, leading to a greater consciousness of quality and costs on the part of domestic manufacturers. The more liberalized technology import policy is helping to bridge the technology gap. All these factors are putting pressures on the machine-tools industry to develop best-practice technology, either by importing or by generating their own. The availability of trained engineers at modest wages has contributed in no small way to the substantial design effort demonstrated by the machine-tools industry. Today, the same engineers can use technical information available in published or unpublished form to find solutions to problems.

The tariff policy in respect of components, raw materials and capital equipment has not been conducive to the generation of further capabilities in these product groups in recent years, as it has favoured the direct import of equipment by providing "negative" effective protection.

The tariff policies followed by the government in respect of the capital goods and machine-tools sector *vis-à-vis* other industry sectors have contributed to a relative lack of profitability in capital goods, leading to sluggishness in the growth in investment in this sector. It has also contributed to a smaller return on capital employed, and smaller profits have also contributed to less effort in R&D and design activities.

In spite of the above factors, the machine-tools industry has been one of the country's leaders in terms of R&D expenditure; this could be attributed to its fight for survival in the environment described above. Government policies have also not contributed any major incentives towards successful R&D efforts.

In the 1980s, the industry developed further, and was able to acquire know-why in machine-tools technology in order to reproduce and even develop new tools, particularly special-purpose machine tools. The human resources necessary for more complex technological development in the 1980s had been created in the earlier period of protection. As a result of the experience with the development of modified versions of machine tools manufactured under licence, the industry had created a pool of technical resources for independent development.

As measures of technological development, the study relied on indicators which could at least approximately reflect learning. For this reason, not only research intensity, but also the employment of engineers and the scope of R&D projects, as well as their duration and cost, were looked at. Of these, data on the scope of R&D projects gave the clearest indication of the extent of self-reliance in the Indian machine-tools industry. Our data show that the large majority of R&D in this industry is directed to exploiting an understanding of know-why in machine-tools technology in order to produce contemporary foreign machine tools. As a reflection of the important role that human resources play in machine-tools development, our data indicate a high rate of employment of engineers in the industry. By contrast, research intensity is modest, and remained unchanged in the last decade. The overall impression that one gets is that the industry has enhanced its technological capability over the previous decade.

To conclude, one can say that the Indian machine-tools industry has shown sustained progress in its acquisition of technological capability. It has created the human resources for further development, subject to market pressures providing the stimulus. In order to understand these transitions in self-reliance in the machine-tools industry, the role of the policy environment, as well as demand and supply conditions, were examined.

With regard to this industry, the Indian government used essentially three policy instruments: (a) the creation of Hindustan Machine Tools (HMT); (b) quantitative controls over imports of machine tools, and restrictions on the purchase of technology until 1983; and (c) the creation of the Central Machine Tools Institute (CMTI). The creation of HMT has proved to be effective in providing a nucleus for the generation of technological capabilities and skills. As already stated, it not only provided the impetus for the machine-tools industry, but also foreshadowed the development of the engineering industry in the country. Quantitative controls were effective for a short period between the mid-1960s and late 1970s, in that the industry could develop modified versions of machine tools manufactured under licence as substitutes for imports. However, these controls insulated the industry from contemporary developments in machine-tools technology, namely the introduction of electronic controls in the mid-1970s, and delayed the process of learning.

The restrictions on technology imports in the late 1960s helped the industry temporarily because it protected indigenous R&D from competition from imported technology. In the early 1980s, these restrictions were a barrier to learning because they deprived the industry of access to modern developments in its field.

We have found no evidence to suggest that CMTI played an important role in design and development. This is because design activity in the machine-tools sector is market-oriented, i.e. associated with specific user requirements, and, hence, is best pursued by manufacturing firms. However, CMTI has facilitated the technological development of the sector by providing high-quality machine and prototype testing services and technical information, and by the creation of human resources.

The changes in the structure demand have also played an important role in the growth of self-reliance in this industry. Between the mid-1960s and the late 1970s, the demand was restricted mainly to general-purpose machine tools. In such an environment, the industry could, at best, acquire the capability to develop modified versions of machine tools for which designs had been purchased under licence. In the

1980s, the demand for machine tools became more varied, and custom-designing has become necessary to meet user needs. In such an environment, there is scope for the development of new machine tools, or at least reproduced versions of foreign machine tools.

Coal-based power-equipment sector
Self-reliance in coal-based power equipment takes place in a context in which large-scale expenditure in scientific research provides a crucial input for design and development. Moreover, engineering design is accompanied by huge expenditures on the testing and quality control of equipment. The scale of R&D expenditure, testing, and investment for the manufacture of large-sized equipment is a major barrier to entry into this industry. Moreover, electrical power equipment is dominated by a few multinational corporations, which enjoy the advantage of long experience in the development and manufacture of such equipment.

Thus, the nature of the industry is such that the absorption of contemporary technology is itself an important goal of self-reliance. The past record of the Indian electrical power-equipment sector shows that it has graduated from the manufacture of small-size units of a maximum of 150 MW in the 1960s to 210 MW in the 1970s and 500 MW in the mid-1980s. In terms of technological capability, the Indian manufactures have reached the second stage of development, i.e. where they are capable of absorbing, adapting, and modifying foreign designs to specific conditions and of undertaking trouble-shooting activities. The transition from the first to the second stage occurred around 1975, when a new corporate plan laid the basis for the formation of a company which could undertake scientific research for modification, adaptation, and trouble-shooting. The role of foreign collaborators became increasingly that of suppliers of basic designs and systems designs. In the 1980s, BHEL has made some initial attempts to graduate to the third stage, i.e. to acquire capabilities for new product development, systems engineering, and project management.

The unit sizes of equipment are relatively small in India. However, the use of modern technology, such as combined cycle cogeneration, has picked up. In fluidized bed combustion technology, India is one of a few countries in the world that have used the technology commercially for boilers up to 30 MW. In MHD also, India has joined a select band of countries able to harness the technology and has already developed a 5 MW (thermal) experimental test facility that has generated power.

In general, self-reliance has reached the stage where the Indian electrical power-equipment industry has acquired the ability to make rational choices of technology and effectively to absorb and adapt them. It has not, as yet, mastered the ability to develop and commercialize new products. This is because of the high costs of development and experimentation in this industry, and because it does not have the benefit of financial instruments to aid indigenous development. As long as the institutional facilities are not available for innovative technological development, one cannot expect higher levels of self-reliance in an industry with high barriers to entry created by high levels of R&D expenditure. Keeping in view these realities of the power-equipment sector, we can conclude that the Indian industry has achieved self-reliance.

To develop the local power-equipment industry, the government has devised three policy instruments: (a) creation of public sector enterprises, i.e. the erstwhile HE(I)L, and BHEL; (b) standardization of unit sizes; and (c) protection. The creation of public enterprises has been of prime importance for the development of the industry in India; but for them, the country would have had little chance of becoming a significant producer of large-sized power-generating equipment because of the high entry barriers facing the industry. The standardization of unit sizes of equipment protected local manufacturers and created conditions for the absorption of imported technology. Finally, protection from imports accorded until 1978 helped the domestic industry to develop without undue external pressures, such as dumping, which, owing to considerable over-capacity in the industry, has been commonly practised by multinational corporations (MNCs) to capture markets. The subsequent liberalization of imports of power equipment partially accelerated the development of self-reliance, although the price-discrimination policies of MNCs halted this process in the mid-1980s.

The factors which have led to increased technological self-reliance in the coal-based power-equipment sector have been threefold. To start with, increased human learning has been a major factor. BHEL's engineers made a concerted effort in the early 1970s to learn from imported machinery installed in India, from the feedback data generated by the equipment installed by them as well as from interactions with foreign consultants who were associated with their foreign buyers. Increased human learning had a compounded effect when these capabilities enabled Indians to purchase superior technologies from the Western world and to learn from them. Secondly, the Indian government policy of standardization of unit sizes permitted the quick absorption of technology.

Thirdly, increased self-reliance was made possible by structural changes in the organization of BHEL, initiated in the mid-1970s, which introduced engineering development centres and a specialized R&D centre. This made it possible for the company management systematically to initiate technology development in the company.

Finally, the need to adapt technology to local raw materials has induced technological development. The high ash content in Indian coal necessitated better combustion methods to extract the maximum thermal energy. Since BHEL's foreign collaborators did not face this situation, the development of indigenous coal-combustion technologies was undertaken in the mid-1970s. This has been one of the most successful areas of R&D within BHEL, and has resulted in the development of fluidized bed boilers, the direct ignition of pulverized coals, and hot gas clean-up systems, as well as many other modifications in designs.

Similar need-based developments in the steam turbine area also resulted in the completely indigenous design and manufacture of 18 MW FDTR turbines.

Process industries

In process-based industries, as in the engineering industries, four successive stages in technological development are delineated: (a) ability to operate and maintain plants commissioned under turnkey contracts; (b) capability to undertake detailed engineering using a basic engineering package supplied by foreign licensors and to undertake project management; (c) capability to design a basic process package using a rudimentary flow sketch and catalyst information; and (d) the capability to develop new process technology or products. However, process industries, as opposed to engineering industries, are characterized by a greater degree of integration and interdependence between components, tailor-made design at a larger level of aggregation, fewer repetitions within a generation for a given growth in the country, a lower mortality rate, and, hence, slower penetration of newer technologies. These characteristics have certain implications for S&T self-reliance, as will be seen below.

Petroleum refining
From a modest beginning with the meagre capacity of 0.25 million tonnes per annum when planning began, the Indian petroleum-refining

industry has come of age with an annual capacity of 45.55 million tonnes at the end of the sixth Five-Year Plan. The rapid expansion of refining capacity has enabled the country to achieve a considerable degree of self-sufficiency in petroleum products and has encouraged the creation of fertilizer, petrochemical, and tertiary downstream industries.

The Indian petroleum industry, which started in the mid-1950s with almost total dependence on foreign sources for know-how, plant and equipment, management, and operational skills, has achieved a considerable degree of indigenization in process know-how, detailed engineering, fabrication of plant, equipment, and manpower. This has been made possible because of the accumulation of technological learning and capability by the industry. Of the four successive stages of technological capability-building in process industries, India has acquired self-reliance in respect of plant operation and maintenance and a substantially complete capability in detailed engineering, equipment fabrication, and project management. In respect of basic engineering, India has acquired process design capability in a number of processes, such as crude and vacuum distillation, Arnine treatment, visbreaking, and dewaxing, which have been put into operation successfully. Thus the industry has reached the stage in technological development where a complete plant can be set up with minimum information from process licensors. In some areas, where India's requirements are unique and where technological needs are repetitive, some innovation and process development activity has been started and pilot plants have been set up.

In acquisition of the technological capability, government policy has played a key role. In accordance with the Industrial Policy Resolution, 1956, which sought public control of industries of basic and strategic interest, the expansion of the Indian petroleum-refining industry took place in the public sector, which also took over the units owned by multinational oil companies in the 1970s. To facilitate technological self-reliance, the government created engineering, design and R&D institutions such as Engineers India Ltd (EIL) and the Indian Institute of Petroleum (IIP).

In-house R&D centres were also set up in major public sector enterprises active in the area. To ensure effective coordination and planning, a number of other institutions were created, such as the Technical Development Committee, the Petroleum Process Development Coordination Group, and the Scientific Advisory Committee of the Ministry of Petroleum. The effective coordination among public sector oil companies, design and engineering organizations (EIL), national laboratories (IIP and NCL), and in-house R&D units has played an important role in bringing about the present degree of technological self-reliance in the Indian petroleum-refining sector.

In addition, the rapid acquisition of technological self-reliance in the industry was facilitated by the relatively mature and stable nature of technology. Furthermore, in a process industry, the nature of links between the suppliers of process designs and special equipment fabricators has important consequences for the achievement of technological self-reliance. Negligible links between suppliers of process and equipment in the international industry have helped Indian attempts to achieve technological self-reliance. Many component units in refineries use equipment such as fabricated vessels, heat exchangers, fired heaters, pumps and compressors, and a variety of electrical equipment and instruments that are identical to those used by other process industries and thermal power plants. The refinery industry could make use of the existing facilities in the country for fabrication of this equipment, though development of these facilities to serve the needs of the refining sector alone would not, perhaps, be viable. Finally, the need to develop site-specific technologies, such as processes suited to crudes recovered in remote locations, has given impetus to the drive for technological self-reliance.

Chemical fertilizers
Like petroleum refining, the nitrogenous fertilizer industry in India is also of recent origin, as the bulk of the production facilities started commercial production in the 1970s.

In the period up to the early 1960s, the expansion of the industry took place almost exclusively through turnkey contracts awarded to foreign firms. But by the mid-1960s the Indian design and project engineering companies had reached the second stage of technological development, i.e. they had acquired the capability to construct integrated fertilizer plants, obtaining the minimum technical assistance from abroad for very specialized plant sections or process techniques. In the period following the late 1960s, local firms successfully built integrated fertilizer plants producing 600 tonnes per day (tpd) ammonia and 1,000 tpd urea as prime contractors. Their responsibilities in these projects included detailed engineering, procurement, erection, inspection, and commissioning of plants. The process know-how for ammonia synthesis, carbon dioxide removal, and urea were obtained from foreign firms. These firms were able to effect a high level of indigenization of equipment and foreign exchange savings in these plants. The stage has been reached where basic engineering designs for urea plants and for all but three sections (reforming, carbon dioxide removal, and ammonia synthesis) of ammonia plants, for which foreign licences are required, are locally available. Indigenous technology developments also attempted to substitute conventional petroleum-based feedstocks, which had to be imported, by abundantly available coal and to up-grade the plant size to 900 tpd ammonia and 1,500 tpd urea, in keeping with trends in the global industry. These developments were effected in five subsequent public sector plants built by local companies.

Fertilizer production is heavily dependent upon use of catalysts in almost every section. The catalysts used in chemical process industries are proprietary items with narrow markets. The development of indigenous capability by local firms to manufacture more stable catalysts is, therefore, an important aspect of the S&T self-reliance achieved by the country.

The impetus for local capability-building in the fertilizer sector was provided by the government in 1963 when it decided to strengthen the technological wing of the Fertilizer Corporation of India (FCI), in order that it might acquire the know-how, both of process and of design and fabrication, in the context of the expansion of fertilizer production capacity in the plans. The technology wing (later called Planning and Development Division) of FCI, which later grew into a separate company (PDIL), along with the Engineering and Development Division of another public sector corporation, Fertilizers and Chemicals, Tranvancore (i.e. FEDO), have spearheaded indigenous technology development in the sector. The technology absorption and indigenization was helped by the replication of plants of the same size. An important impulse for technological development came from the need to substitute imported feedstock by local coal in order to save foreign exchange. The proprietary and oligopolistic hold over the supply of catalysts also led local firms to develop local technology for their manufacture.

The study of the fertilizer industry also highlights the role that finance plays in technology selection. It has been shown that reliance on foreign financial resources has prevented the fuller utilization of local technological capabilities, especially in the more recent period, when the government decided on 1,350 tpd as the standard size for further plants in preference to 900 tpd, for which local technology was available.

Factors in technological development

From the analysis of S&T self-reliance at the overall level and the industry studies, a number of policy and structural factors in technological development emerge.

Policy factors

Import substitution
Industrialization itself is an important source of accumulation of technological learning. In this sense, the import strategy of the Indian government, which fostered the development of a wide range of industries, is particularly important, because the presence of these industries facilitated the unpackaging of technology imports, and hence helped absorption.

Human resource development and S&T infrastructure
The expansion of infrastructure for technical and higher education under the Scientific Policy Resolution, 1958, which ensured an adequate supply of qualified S&T personnel, has been of great value for S&T self-reliance. It has facilitated the quick replacement of foreign personnel and absorption of imported technology. In addition, the network of national laboratories has proved to be a major source of expertise and of other technical services such as testing, standards, and technical information. The role of national laboratories in designing and innovations, however, varies from industry to industry. In petroleum refining, IIP in collaboration with EIL appears to have productionized a number of processes; CMTI, in machine tools, does not appear to have played an important role in designing. Though the observations made here do not warrant generalization, the two determinants of success of national laboratories appear to be the nature and extent of laboratory–industry interaction and the extent of market orientation of products. The more extensive the laboratory–industry linkage (as in petroleum refining), the greater is the likelihood that the laboratory will be an important source of innovations. And the greater the market orientation of the product (as in the case of machine tools, particularly the custom-made ones), the less is the chance of a laboratory being a successful innovator.

Direct intervention

In all the industries studied, the public sector enterprises – i.e. HMT and BHEL in the engineering industries and CEDOs such as EIL, PDIL and FEDO in process industries – emerge to be the nuclei for technological development. This is particularly significant, because in all these areas, except machine tools, there were high entry barriers for innovation, as a result of a large minimum scale of R&D activity. Public sector industrial enterprises, because of the relatively large scale of their operations, were able to finance and coordinate the requisite level of technological activity.

Protection of indigenous technology

The industry studies uphold the view that local technology development is like rearing an infant: the industry requires protection and support in the initial period, but finally grows up.[51] In these areas, effective protection to local technology, at least until the late 1970s, facilitated the local ownership of user industries, as in power generation, petroleum refining, and fertilizers; and in the case of machine tools, this was achieved by quantitive controls over imports. The existing technology import regulations alone could not guarantee protection to local technology in general. For instance, in fertilizers almost all private sector plants were built by foreign companies on a turnkey basis, even after local design and plant fabrication capabilities became available. In this context, we have observed the inadequacy of general technology import policies in providing protection; in order to spur the generation and utilization of technology, the need to supplement the "stop–go" type of import regulation by price protection for local technology has been emphasized.

Standardization of unit sizes

For process industries, the choice of unit size has an important bearing on the development of local technological capability. Standardization of unit sizes by the government in the case of power equipment, petroleum refining, and fertilizers has helped rapid absorption and mastery of technologies because it has made possible the frequent replication of similar plants.

Structural and industry-specific factors

Technological maturity and pace of technological change

S&T self-reliance is achieved more easily in industries with relatively mature and stable technologies, such as the process industries, than in those undergoing rapid technological change.

Nature of international technologies markets
The nature of international markets, in respect of the seller concentration and the degree of vertical integration in an industry, affects national attempts to achieve S&T self-reliance. If the market of a particular technology is particularly oligopolistic, the technology may not be available in the desired mode, such as on a licensing basis. The choice of the mode of technology import has been found to influence local technological capability-building. Secondly, the degree of vertical integration of the technology suppliers or the nature of links between process (or design) suppliers and equipment suppliers also affects the attempt to achieve S&T self-reliance. Extensive links between process and equipment suppliers deny the local engineering industry the chance to fabricate equipment, and hence hinder attempts towards unpackaging the technology imports.

Location-specific technological needs
The need to develop or adapt the existing technology to suit local conditions or requirements turns out to be an important factor facilitating self-reliance. Our case-studies show that the need to adapt technology to Indian coal with its high ash content (in the case of the thermal power-equipment sector), the need to develop processes specifically suited to crudes recovered in remote locations (in the case of the petroleum-refining industry), and the need to replace the petroleum-based feedstock by coal (in the case of fertilizers) have provided important stimuli to the development of local technology in these areas.

Organizational structure
The organizational structure of the enterprise can also have important consequences for technological development, as our study of the power-equipment industry demonstrates.

Concluding remarks

The study has been conducted in two stages, the one dealing with the overall economy level and the second analysing the process of technological development in four industries, two of which are engineering (machine tools and thermal power equipment) and two process-based industries (petroleum refining and chemical fertilizers). All the industries selected have enjoyed a key place in India's plans for import sub-stitution and industrialization because of the intensive linkages they have with other sectors. The analysis, both at the overall macro level and in the industry case-studies, has attempted to assess the level of S&T self-reliance achieved, and to bring out the role of government policies and other factors in facilitating technological transformation.

India's planning for self-reliance in S&T has sought to reduce the country's dependence on imported S&T resources. In order to assess whether dependence has actually declined, the study analysed trends in a number of indicators proxying different aspects of dependence on foreign technology, charting the build-up of an autonomous capability to absorb, adapt, and indigenize imported technology, and to innovate and develop products and processes locally. The trends observed in each of these indicators reveal considerable progress towards the achievement of S&T self-reliance. In respect of the availability of skilled and technical manpower for operating plants and of design, engineering, and fabricating equipment, India has achieved almost total self-sufficiency, particularly in industries in which the rate of tech-nological change is not fast.

Whether judged in terms of S&T infrastructure and other indica-tors of S&T development or in terms of performance (e.g. technology exports), the intercountry comparisons have grouped India with a few countries that have the most advanced S&T capabilities in the developing world.

The achievement of India in the sphere of S&T capability-building, though commendable, was found to fall short of expectations and potential. In the light of the determinants of innovative activity and utilization of local technology *vis-à-vis* imported ones, some directions for future technology policy have also been outlined.

The development of S&T capabilities is a complex process that is influenced by a wide array of social, cultural, economic, and external factors. This study has brought out a number of factors found to affect overall S&T development and technological capability-building in two engineering and two process industries in India. The findings suggest that public policy and direct governmental intervention have played a central role in capability-building. As there is considerable inter-industry variation in their relative roles, a fuller comprehension of the factors contributing to the technological development would necessitate further analysis of more industries, particularly those with different ownership characteristics and market structures, and growth profiles other than the ones studied here. Similarly, intercountry analyses may

bring out the role of sociocultural factors in technological development. Nevertheless, it is hoped that the present study will lead to more
extensive research on this subject of vital importance, and that it will
prove useful in providing a conceptual framework for further work.

Notes

1. P.C. Mahalanobis, *Talks on Planning*, Calcutta: Indian Statistical Institute, 1961, especially pp. 69–70.
2. I.e. the fourth (1969–1974), fifth (1974–1979), and sixth (1980–1985) Five-Year Plans. Planning Commission, *Five Year Plans*.
3. Vijay L. Kelkar, "India in the World Economy, Search for Self-reliance," *Economic and Political Weekly*, annual no., February 1980.
4. Central Statistical Organisation, *National Accounts Statistics*, New Delhi, 1987.
5. V.L. Kelkar and K.M. Raipuria, *Manufactures Exports in India's Development: Policy Options for the Eighties*, Tilburgo, Netherlands: IDPAD, 1985.
6. Planning Commission, *Sixth Five Year Plan, 1980–85*, and C. Rangarajan, "Seventh Plan Industrial Prospects and Opportunities," *Economic Times*, 6 May 1986.
7. For a recent discussion, see V.R. Panchamukhi and K.M. Raipuria, "Productivity and Development Strategy in India: Major Dimensions, Issues and Trends 1965/66 to 1982/83," RIS Monograph no. 2, 1985; revised version, March 1986.
8. See note 7 above.
9. Planning Commission, *The Seventh Five Year Plan*, vol. 1, p. 10.
10. Planning Commission, *The Third Five Year Plan, 1961–66*, p. 107.
11. Planning Commission, *The Sixth Five Year Plan, 1980–85*, p. 320.
12. See note 11 above, and Planning Commission, *The Fifth Year Plan, 1974–79*, p. 231.
13. It is proposed to adopt 'zero-based budgeting' in regard to total Plan and non-Plan outlay for S&T in the seventh Plan.
14. This discussion draws on Nagesh Kumar, "Technology Policy In India: An Overview of Its Evolution and an Assessment," in P.R. Brahmananda and V.R. Panchamukhi, eds., *The Development Process of Indian Economy*, Bombay: Himalaya, 1987, pp. 461–492.
15. See Ministry of Industry, *Guidelines for Industries, Part I, Policy and Procedures*, New Delhi: Indian Investment Centre, 1982, chap. 4.
16. DSIR, *Annual Report 1985/86*, New Delhi: Govt of India, Department of Scientific and Industrial Research, 1986.
17. Ashok V. Desai, "The Origin and Direction of Industrial R&D in India," *Research Policy* 9 (1980):85.
18. Indian Investment Centre, *Technologies from India*, New Delhi: IIC, 1982.
19. See note 16 above.
20. For instance, see Desai (note 17 above) and Nagesh Kumar, "Cost of Technology Imports: The Indian Experience," *Economic and Political Weekly* 20, 31 August 1985, pp. M-103–114.
21. See, for instance, *India Today*, 15 June 1984.
22. *The Long Term Fiscal Policy Statement*, 1985.
23. *India 1984: A Reference Annual*, New Delhi: Government of India, Publication Division, 1989, p. 84.
24. For example, J.C. Ghosh, Presidential Address of 1943 to National Institute of Sciences in India, Madras, quoted by Mahalanobis (note 1 above).
25. Department of Scientific and Industrial Research, *Annual Report 1985–86*, p. 28.
26. *India: Non-electrical Industrial Machinery Manufacturing – A Subsector Study*, document of the World Bank, Report no. 5095-IN, 2 August 1984, reviewed in *Economic and Political Weekly* 20, 12 October 1985, pp. 1724–1725.

27. See note 26 above.
28. This study was commissioned by the Confederation of (Indian) Engineering Industry (CEI) and has been cited by K.C. Khanna, "Why Industry Stagnates," *Times of India*, 28 October 1986.
29. See note 26 above.
30. See, for instance, Ashok Desai, "The Origin and Direction of Industrial R&D in India," *Research Policy* 9 (1980): 74–96.
31. See the unpublished interviews conducted by Carlos Cordeiro and reported in Louis T. Wells, Jr, *Third World Multinationals: The Rise of Foreign Investment from Developing Countries*, Cambridge, Mass.: MIT Press, 1983.
32. See review of studies in Nagesh Kumar, "Foreign Direct Investments and Technology Transfers among Developing Countries," in V.R. Panchamukhi et al., eds., *The Third World and the World Economic System*, New Delhi: Radiant, 1986, pp. 139–165.
33. Richard Thomas, *India's Emergence as an Industrial Power*, New Delhi: Vikas, 1982.
34. See Donald J. Lecraw, "Direct Investment by Firms from Less Developed Countries," *Oxford Economic Papers*, vol. 29, 1977, pp. 442–457.
35. Revealed by M.M. Mehta of Tractor Manufactures Association at the National Seminar on In-house R&D in Industry, organized by DSIR and CEI, New Delhi, April 1987, and reported in Sunil Mani, "Small Sector Scores in In-house R and D," *Economic and Political Weekly*, 18 July 1987, pp. 1174–1176.
36. Revealed in different presentations at the National Seminar reported in note 35.
37. A.S. Bhalla, "Can High Technology Help Third World Take-Off?" *Economic and Political Weekly*, 4 July 1987, pp. 1082–1085.
38. Norbert Konrad and Dietrich Wahl, "Scientific and Technological Potentials in Developing Countries as National and International Economic Factors," *Economic Quarterly* 20 (1985), no. 3: 3–18.
39. See note 38 above.
40. Sanjaya Lall, "Trade in Technology by a Slowly Industrialising Country: India," *Multinationals, Technology and Exports Selected Papers*, London: Macmillan, 1985, pp. 203–226.
41. See note 40 above.
42. For instance, the gross income of NRDC, which is the sole agency for licensing CSIR innovations, from royalty and premia on account of licensing in the year 1981/82 was a meagre Rs. 10.2 million, compared to remittances of Rs. 2,866.9 million sent abroad by enterprises in royalty and technical know-how fees in the same year and an expenditure of nearly Rs. 100 crores in CSIR laboratories. See Department of Science and Technology, *Research and Development Statistics, 1980–81*, New Delhi: DST, 1982.
43. See Desai (note 17 above) and G. Alam and J. Langrish, "Government Research and its Utilisation by Industry: The Case of Industrial Civil Research in India," *Research Policy*, January 1984.
44. Department of Scientific and Industrial Research, *Annual Report 1985/86*.
45. The following discussion draws on Nagesh Kumar, "Technology Policy in India: An Overview of its Evolution and an Assessment," in P.R. Brahmananda and V.R. Panchamukhi, eds., *The Development Process of the Indian Economy*, Bombay: Himalaya, 1987, pp. 461–492.
46. See Amiya Kumar Bagchi, "Public Sector Industry and Quest for Self-reliance in India," *Economic and Political Weekly*, special number, 1982, pp. 615–628.
47. See Morten I. Kamien and Nancy Schwartz, *Market Structures and Innovation*, Cambridge: Cambridge University Press, 1982, for a survey of the literature.
48. See Ashok V. Desai, *Market Structure and Technology: Their Independence in Indian Industry*, WEP Working Paper no. 117, Geneva: International Labour Office, 1983; and Nagesh Kumar, "Technology Imports and Local Research and Development in Indian Manufacturing," *The Developing Economies* 25 (1987), no. 3, for empirical evidence. Katrak, in his empirical study of Indian manufacturing, also finds that larger enterprises undertake pro-

portionately less R&D than smaller ones: see Homi Katrak, "Imported Technology, Enterprise Size and R&D in a Newly Industrialising Country: The Indian Experience," *Oxford Bulletin of Economics and Statistics* 47 (1985): 213–229.

49. See Kumar (note 45 above) for an elaboration of this point.

50. See Kumar (note 48 above) and K.K. Subrahmanian, "Towards Technological Self-reliance: An Assessment of Indian Strategy and Achievement in Industry," in P.R. Brahmananda and V.R. Panchamukhi, eds. (note 45 above), pp. 420–446, for evidence at industry level. DGTD studies cited in P. Mohanan Pillai, "Technology Transfer, Adaptation and Assimilation," *Economic and Political Weekly*, November 1979, pp. M-121–126, and UNCTAD, *Technology Issues in the Capital Goods Sector: A Case Study of Leading Machinery Producers in India*, Geneva: UNCTAD, 1983, provide evidence at the firm level.

51. An earlier study of the Indian tractor industry also reached a similar conclusion: see Ward Morehouse, "Technology and Enterprise Performance in the Indian Tractor Industry: Does Self-reliance Measure Up?," *Economic and Political Weekly*, 20 December 1980, pp. 2139–2152.

2

China

The research objective of this study is to present China's experience in attempting to meet its target of "self-reliance." The research was carried out at both a macro and a micro level. Special attention is given to S&T policy, which is centred around the following: (a) the role of government; (b) the socio-economic background to S&T policy; (c) planning and coordination of S&T policy with respect to socio-economic development; and (d) policies to encourage the efficient use of inputs to the S&T system, e.g. financial and human resources.

The Chinese study used a systems approach. The S&T system, economic system, social system and international system were studied as interrelated parts of an integrated whole. The system is a dynamic one, which changes with geographical location and time (fig. 1). We started by reviewing the past, and moved on to analysing the past and current status of S&T and its prospects for the future (fig. 2). The S&T system was analysed according to a life-cycle concept (fig. 3).

Historical perspective

China has carried out a socialist transformation and construction of the economy since 1949. The characteristic features of the Chinese socio-economic system can be defined with reference to such factors as ownership, decision-making, motivation, information, and coordination structure. The patterns of past socio-economic development in terms of these basic features are described below.

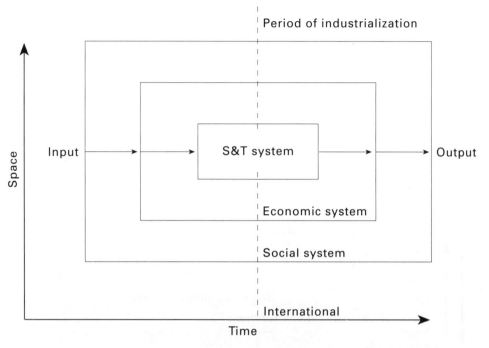

Fig. 1. System in time and space phase

The completion of basic socialist transformation, 1949–1956

Following the three years of economic rehabilitation from 1949 to 1952, China initiated and established a relatively flexible central planning system. The first Five-Year Plan for national economic development was carried out between 1953 and 1957. An initial basis for socialist industrialization, involving the construction of 694 "above-norm" projects (including 153 major ones), was planned and completed.

Between 1953 and 1956, the annual average increase in the gross output value of industry was 19.6 per cent and of agriculture 4.8 per cent. During that period, more than 110 large industrial enterprises were completed, mostly in heavy industry. This laid the groundwork for Chinese socialist industrialization. The value of the industrial output of the state-owned enterprises reached around 53 per cent, and that of the collectively owned 19 per cent, in the year 1957. The remaining industrial enterprises were in the category of either joint state/private ownership or private ownership.

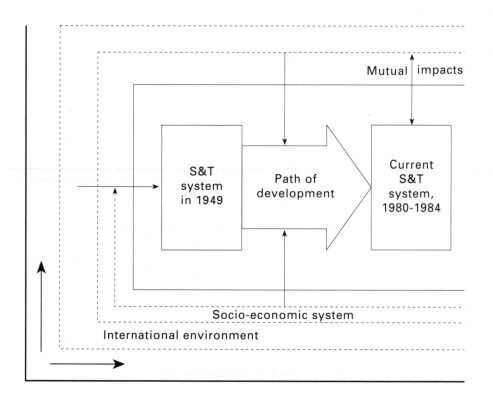

Fig. 2. Research methodology 2

Planning in that period was relatively flexible, the scope for control was limited to minor parts of the state-owned enterprises, important materials, and projects. Indirect control of the cooperative, individual, private capitalist and state capitalist economy was through proper economic policies, the pricing system, taxation, and the credit system.

In the agriculture sector, the Agrarian Reform Law of the PRC was promulgated in June 1950. After the land reform, the agricultural production system was changed in three stages: the mutual aid team; the elementary agricultural producers' cooperative; and the advanced agricultural producers' cooperative.

Pattern of socio-economic development from 1956 to 1977

China undertook full-scale socialist construction in the period from 1956 to 1966. Between 1956 and 1966, fixed assets in industry grew four times in value while national income increased by 58 per cent in

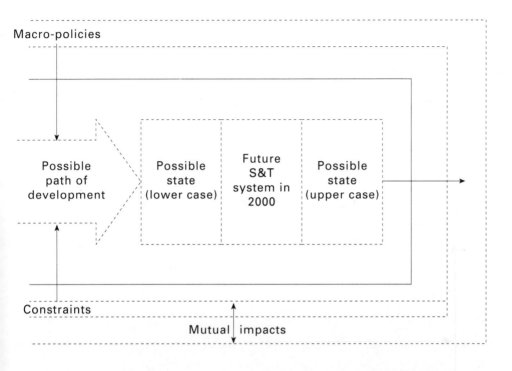

terms of comparable prices and by 34 per cent in terms of per capita amount. From 1966 to 1976, the ten-year Cultural Revolution brought China its biggest set-back since 1949; but, in spite of this, China as a whole still had a relatively high rate of economic growth in that period. The grain output rose from 193 million tons in 1956 to 282.73 million tons in 1977. Also in this period, crude steel output rose from 4.45 million to 23.74 million tons, coal from 11 million to 93.6 million tons, chemical fertilizers from 0.133 million to 7.238 million tons, machine tools from 25,928 to 198,700 sets, and cotton cloth from 5.770 billion to 10.151 billion metres.

The aim of the ownership system of that period was to strengthen the state-owned economic system and weaken the collectively owned system. Private ownership had almost been abolished. In rural areas, the rural commune economy was established. In the commercial sector, the sales of state-owned enterprises reached over 90 per cent during this period. The commodity circulation system suffered considerably.

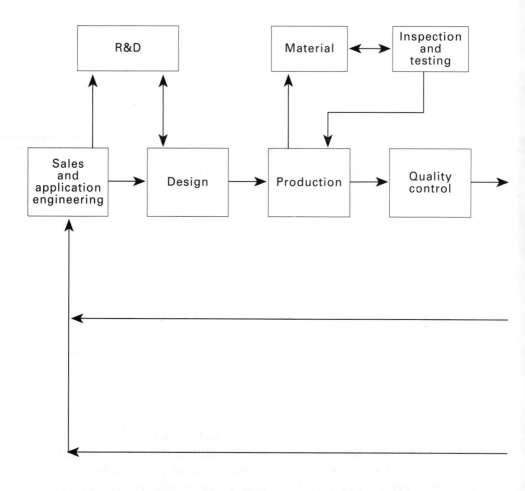

Fig. 3. Complete life cycle and component stages of S&T system: research methodology 3

The ambit of mandatory planning in the coordination system was increased after 1956, while in the period of the "Great Leap Forward" (1958–1960), planning control was decentralized to the provincial level. Owing to the lack of effective macro-control, the economic system did not function efficiently. Central planning was once again emphasized in the adjustment period of 1963–1965, but, because the planning system was again disrupted in the period of the Cultural Revolution, there were no improvements or modifications in the conception or

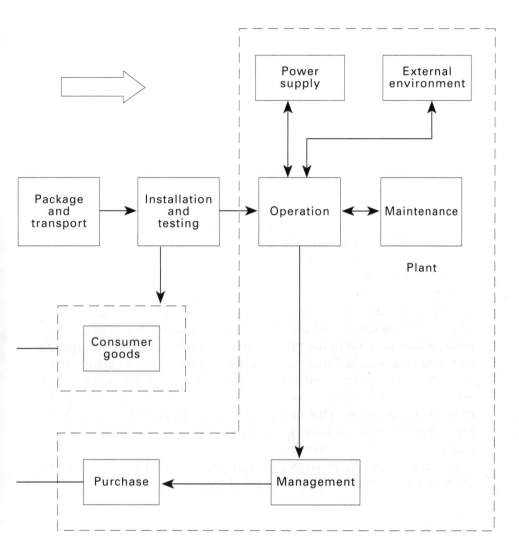

practice of planning at that time. In particular, the information flow was weakened.

The new pattern of socio-economic development after 1978

Since 1977 there has been intense discussion in China about the pattern of economic development. Reform was initiated first in the rural areas by the introduction of the "production responsibility system."

The system of contracted responsibility for production, with remuneration linked to output, greatly strengthened the performance of the agricultural sector. Furthermore, a variety of rural activities was encouraged, especially the development of rural enterprises. In 1985, the gross value of the industrial output of rural enterprises ran at around 30 per cent of the gross industrial output of the whole nation.

China pressed ahead with its rural reforms, and the rural economy moved towards specialization, large-scale commodity production, and modernization. As a result of the success of this reform, and of a series of experimental reforms in selected areas of the urban economy, a document on the "Decision on Reform of the Economic Structure" was adopted by the Twelfth Central Committee of the Communist Party of China at its Third Plenary Session. Emphasis was placed on the fact that "invigorating enterprises is the key to restructuring the national economy."

Stress was also placed on removing obstacles in the way of the development of the collective economy and the individual economy in cities and rural towns and the creation of conditions for their development, which gave them the protection of the law. The scope of mandatory planning was reduced and guidance planning was extended step by step. Certain farm and sideline products, small articles for daily use, and labour services in the service and repair trades were subjected to market regulation. This new pattern of socio-economic growth was rapid. For example, in 1985, grain production was 370 million tons and coal production 620 million tons.

China's development strategies and policies from the 1960s to the 1980s were directed toward two main objectives:

Industrialization

Before the 1980s, the importance of developing a heavy industrial base was particularly emphasized. Before 1977, this development effort resulted in the creation of almost the entire range of modern industries. China was now nearly self-sufficient in those industries making capital equipment. This objective was achieved through high investment expenditure and a massive infusion of centrally mobilized resources. Consequently, consumption grew more slowly than income. But, after the economic reforms of 1977, government policies on consumption aimed not only at ensuring a more rapid improvement in living standards, but also at narrowing the differential between heavy and light industry.

Poverty reduction

The second major development objective was the elimination of the worst aspects of poverty. Owing to the past pattern of development, there was a general absence of individual incomes from property. The income share of the richer group was small; and as a fundamental policy objective the low-income groups had their basic needs satisfied. Formerly the food supply was also guaranteed through a mixture of state rationing and collective self-insurance. The great majority also had access to basic health care, education, and family-planning services. Consequently, life expectancy was raised from 37.7 in 1960–1965 to 64.2 in 1975–1982. The life expectancy at birth of males and females in 1984 was respectively 68 and 70.

The main economic policies in the 1980s and the relationship between the main social and economic policies and the S&T policies are listed qualitatively in table 1.

Figure 4 shows the pattern of growth of the Chinese economy by sector from the 1950s to the 1980s, as well as the trends after 1978.

The growth rates for the main agricultural and industrial outputs are also shown in table 1. The figure clearly demonstrates the socio-economic development pattern and its impact on the agricultural and industrial sectors.

The growth rate in the agricultural sector from the 1960s to 1977 was relatively slow, owing to the lack of scope for the expansion of cultivated land and also to excessive governmental intervention in the application of the misguided "grain first" policy. From the figures for the rate of growth of grain, cotton, etc., in table 1, the effects of the socio-economic development pattern and policy are self-explanatory. The rate of multiple cropping had been increased to an average of 1.5, and traditional labour-intensive cultivation techniques had been refined by extensive improvements in modern irrigation (table 2).

Since the Second Five-Year Plan emphasized the government's policy of promoting the growth of industry, the amount of inputs – capital, labour, and materials – was greatly increased with respect to the output. The growth rate of the industrial sector was consequently high between the 1960s and 1977.

The status and role of S&T at various stages of development is demonstrated by the following three factors:
1. The growth of S&T personnel and the education system (tables 3 and 4).

Table 1. Output of selected agricultural and industrial sectors showing the effect of socio-economic development pattern and policies

	1960	1964	Growth rate (%)	1965	1977	Growth rate (%)	1978	1985	Growth rate (%)
Agricultural sector									
Grain (10^4 tons)	14,350	18,750	6.91	19,453	28,273	3.16	30,477	37,911	3.17
Cotton (10^4 tons)	106.3	166.3	11.84	209.8	204.9	−0.20	216.7	414.7	9.72
Oil-bearing crops	194.1	336.8	14.77	362.5	401.7	0.86	521.8	1,578.4	17.13
Industrial sector									
Cotton cloth (10^8 m)	54.5	47.1	−3.58	62.8	101.5	4.08	110.3	146.7	4.16
Crude steel (10^4 tons)	1,866	964	−15.22	1,223	2,374	5.68	3,178	4,679	5.68
Chemical fertilizers (10^4 tons)	40.5	100.8	25.60	172.6	723.8	12.69	869.3	1,322.2	6.17
Crude oil (10^4 tons)	520	848	13.01	1,131	9,364	19.26	10,405	12,409	2.64
Coal (10^8 tons)	3.97	2.15	−14.21	2.32	5.50	7.46	6.18	8.72	5.04
Electricity generation (10^8 kWh)	594	560	−1.46	676	2,234	10.47	2,566	4,107	6.95
Bicycles (10^4)	176.5	170.5	−0.86	183.8	742.7	12.34	854.0	3,227.7	20.92
Television sets (10^4)	0.79	0.21	−28.20	0.44	28.46	41.55	51.73	1,667.66	64.24

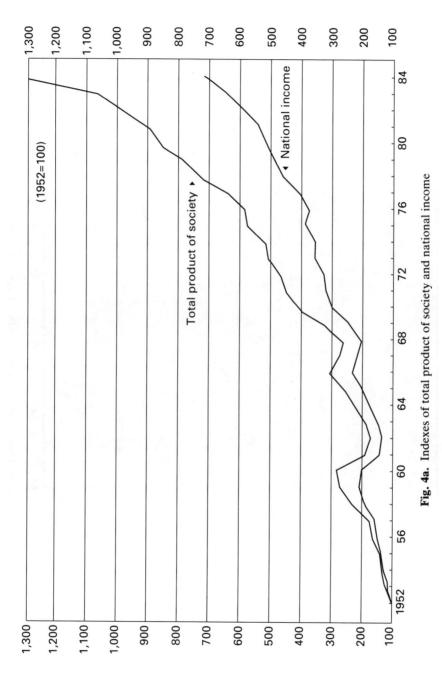

Fig. 4a. Indexes of total product of society and national income

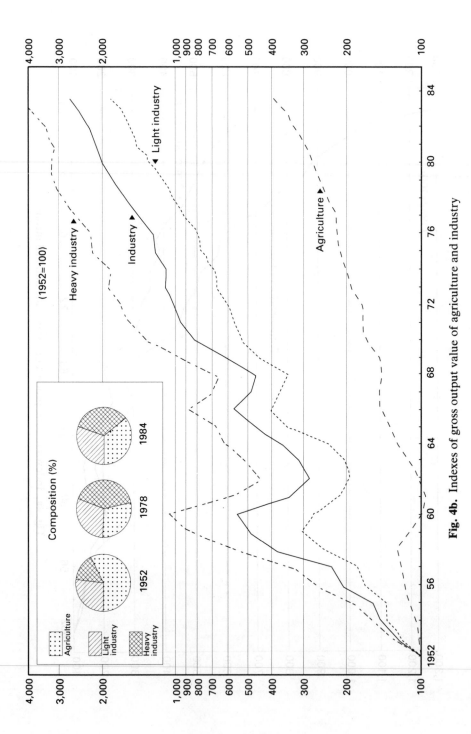

Fig. 4b. Indexes of gross output value of agriculture and industry

Table 2. Degree of mechanization, electrification, farmland under irrigation, and amount of chemical fertilizer used in Chinese agricultural sector

Item	1952	1957	1965	1977	1980
Mechanically tilled land area					
In (10^4 ha)	13.6	263.6	1,557.9	3,841.0	4,099.0
% of total arable land (%)	0.1	2.4	15.0	38.7	41.3
Irrigated area (10^4 ha)	1,945.9	2,733.9	3,305.5	4,499.9	4,488.8
% of arable land	18.5	24.4	31.9	45.3	45
with mechanical and					
electrical irrigation	31.7	120.2	809.3	2,434.9	2,523.1
Electricity consumption					
in rural area (10^8 kWh)	0.5	1.4	37.1	221.9	320.8
Chemical fertilizers					
utilized	7.8	37.3	194.2	648.0	1,269.4

2. Major S&T events from the 1960s to the 1980s:
– October 1964: The first successful test of an atomic bomb.
– September 1965: Synthesis of insulin.
– October 1966: Successful guided atomic missile test.
– 1966–1967: Comprehensive survey team comprising 30 different disciplines carries out systematic survey in regions above 7,000 metres on Mount Qomalangma.
– June 1967: Test of H-bomb.
– April 1970: First Chinese earth satellite (173 kg in weight) launched.
– September to December 1979: Optical fibre communication system established and tested for operation in Shanghai, Beijing, and Wuhan.
– October 1979: Liver cancer diagnosis at an early stage through radio-rocket electrophoresis with autoradiography.
3. Programme effort. This was one of the Chinese S&T successes directed by the central government, in which resources were concentrated on large specific projects. However, for a large variety of products this had to be supplemented by the technology market.

National factor endowments

Natural material resources

The basic physical resources of China – natural resources and existing production capabilities for important products – are shown in tables 5 and 6.

Table 3. Natural scientific and technological personnel in state-owned units (unit: 10^4 persons)

Item	1952		1978		1980		1981		1982		1983		1984		1985	
Engineering	16.4	(38.6)[a]	157.1	(36.1)	186.2	(35.3)	207.7	(36.4)	235.4	(37.6)	280.2	(40.9)	316.2	(42.4)	340.4	(43.6)
Agriculture	1.5	(3.5)	29.4	(6.8)	31.1	(5.9)	32.8	(5.7)	36.2	(5.8)	40.5	(5.9)	43.5	(5.8)	45.1	(5.8)
Public health	12.6	(29.6)	127.6	(29.4)	153.3	(29.0)	168.0	(29.4)	180.7	(28.8)	193.4	(28.2)	207.8	(27.8)	216.1	(27.6)
Scientific research	0.8	(1.9)	31.8	(7.1)	32.3	(6.1)	33.8	(5.9)	37.2	(5.9)	32.8	(4.8)	33.5	(4.5)	33.6	(4.3)
Teaching	11.2	(26.4)	89.4	(20.6)	125.0	(23.7)	129.1	(22.6)	136.9	(21.9)	138.3	(20.2)	145.6	(19.5)	146.5	(18.7)
Total	42.5		434.5		527.6		571.4		626.4		685.2		746.6		781.7	
Technical personnel per 10,000 population (individuals)	7.4		45.7		53.7		57.4		62.0		67.1		72.5		74.7	
Technical personnel staff and workers per 10,000 population (individuals)	269.0		593.3		657.9		682.5		725.8		781.2		864.4		869.5	

a. Figure in parentheses are percentages of total.

Table 4. Development of education system

| | Number of schools | | | Number of students (in tens of thousands) | | |
	1949	1980	Factor increase of 1980 over 1949	1949	1980	Factor increase of 1980 over 1949
Institutions of higher learning	205	675	3.3	11.7	114.37	9.8
Secondary specialized schools	1,171	3,069	2.6	22.9	124.34	5.4
Ordinary middle schools	4,045	118,377	29	103	5,508.08	53.5
Primary schools	346,800	917,316	2.6	2,439	1,462.96	6
Kindergartens	1,300	170,419	131	1.3	1,150.77	88.5

Table 5. Feature of some selected Chinese natural resources

Item	Unit	China	India	Japan	World average
Arable land	Ha	115–133 million	173		–
Arable land	Ha/capita	0.1	0.2	0.03	0.35
Forest covering	%	12	22.7	67.9	31.3
Grassland	Ha	224 million	11 million	0.6 million	–
Annual stream flow (water resources)	m^3	2,614.4 billion	–	–	47,000 billion
Annual stream flow per capita	m^3/capita	2,563	–	–	
Hydropower resources	10 million kW	6.76	–	–	10,800
Proved coal reserves	10^9 tons	737.1[a]	121.36[a]	–	–
Iron ore reserves	10^9 tons	47.20	22.4	–	–

a. Source: *Statistical Yearbook of China 1985.*

Table 6a. Output of main industrial products

	Output in 1980	Output in 1985	Factor increase of 1980 over 1949
Coal (10^6 tons)	620	872	18.4
Crude oil (10^6 tons)	105.95	124.90	882
Electricity (10^6 kWh)	300,600	410,700	69
Steel (10^6 tons)	37.12	46.79	231
Machine tools	134,000	121,000	82
Chemical fertilizers (10^6 tons)	12.32	13.22	2,160
Cotton yarn (10^6 tons)	2.93	3.53	8
Sugar (10^6 tons)	2.57	4.51	11.8
Wristwatches (units)	22.67	54.47	
TV sets (units)	2.49	16.67	
Radios (units)	30.04	16.00	

Table 6b. Output of main farm products (10^6 tons)

	Output in 1980	Output in 1985	Factor increase of 1980 over 1949
Grain	320.56	379.11	1.8
Cotton	2.707	4.147	5.1
Edible oil	7.691	15.784	2.0
Pork, mutton, beef	12.055	17.607	4.4
Aquatic products	4.5	7.05	9.0

Although China has a fair amount of natural resources, in per capita terms they are low, and unevenly distributed. A proper locational policy for industry is necessary in order to manage correctly the country's transportation and resource distribution problems. China world ranking for certain products is shown in table 7.

Human resources

China is rich in human resources. Her population was 1.045 billion in 1985. The educational level of the population has improved, but figures on a percentage basis still fall short of those of other countries, particularly the gross enrolment ratio in higher education.

China's factor endowments vary widely between regions. At the initial stage of development, human resources and sociocultural factors

Table 7. World ranking in production capability of selected products in 1985

Item	World ranking
Grain	2
Cotton	1
Meat	2
Steel	4
Coal	2
Crude oil	6
Electricity generation	5
Value added (industrial)	
Value added (agricultural)	1

play a more critical role than natural factor endowments. But factor endowments are also important as a potential source of development.

After 1949, with the emphasis on local self-sufficiency rather than specialization, the comparative advantages of different regions were not fully explored. A comparative study between Shanghai (a formerly relatively developed metropolis) and the North-west region (which was underdeveloped before 1949) was carried out and this illustrates some of the problems.

Shanghai was a large metropolis before 1949. It had begun to develop modern industry in 1865, when the Jiangnan Bureau of Manufacture was established. By the time of the First World War there were more than 100 enterprises. The first electric power plant was constructed in 1882, and the Shanghai–Nanjing and the Shanghai–Hanchow railways were completed in 1908 and 1909 respectively. Banking and financial services were also developed in the early twentieth century. Therefore, although Shanghai had hardly any mineral resources, its industrial output was quite high owing to its skilled labour force, convenient transportation, and sociocultural factors: before 1949, there were around 200,000 privately owned enterprises and a total number of employees of 428,000, constituting respectively 36.01 per cent and 26.06 per cent of the total for the whole nation.

The North-west region of China comprises five provinces: Shannxi, Gansu, Ningxia, Qinghai and Xinjiang. Here, many nationalities live together. Lying on China's north-west border, it forms an arid and semi-arid belt. Its area is vast and its population sparse, but it is resource-rich (table 8). Historically, it was the birthplace of Chinese culture and the place where occidental and oriental civilizations once converged. None the less, communication and transportation in this

Table 8. Natural resources and factor endowments in North-west region, 1985

Item	Shaanxi	Gansu	Ningxia	Qinghai	Xinjiang
Population (10³)	30,020	20,410	4,150	4,070	13,610
Mineral resources					
No. of types	86	66	99	59	115
Major resources	Molybdenum, mercury, asbestos, coal	Nickel, platinum	Coal, gypsum	Lithium, potassium salt, magnesium salt, sulphur, cobalt	Coal, petroleum
Grassland (10³ ha)	13,330			33,450	50,000
Arable land (10³ ha)	4,113	2.26ᵃ	933	588	9,330

a. Ha per capita available.

region are poor, with some places difficult to access. Furthermore, the ecological environment is quite fragile. Although huge strides have been made in economic and social progress since 1949, the region's level of development is comparatively low.

Table 8 shows the natural factor endowments of the North-west region, and table 9 selected data for comparative purposes. It can be seen that the North-west is rich in natural resources, but also that its Gross Value of Industrial Output (GVIAO) is far behind that of Shanghai. Table 8 also shows the economies of scale and comparative advantages of Shanghai and the region, and reveals that different areas (provinces or municipalities) have nearly the same industrial structure. While the efforts made by different provinces and regions towards "self-reliance" are evident, it is also clear that the comparative advantages of the regions have not been fully explored.

Agricultural sector

Through their long experience of farming, the Chinese have developed a whole series of traditional methods with intensive cultivation as the key link. Internal upheavals and warfare in the first half of the twentieth century weakened the limited development of agricultural services and led to the widespread destruction of the rural infrastructure. After liberation, a socialist managerial system was adopted in China's countryside and remuneration by workpoints was popularized. The collective economy of China's agriculture for years was affected by an overconcentration of managerial powers and a unitary form of operation, both of which dampened peasant enthusiasm. But considerable development of irrigation and drainage systems to regulate the water supply, the provision of chemical fertilizers by the government, and the very high input of labour promoted agricultural growth. The rural reform also stimulated exceptionally rapid growth in agricultural production. Rural incomes and food consumption rose as a result of rapidly rising yields.

Industrial sector

Nearly the entire range of modern industry has now been set up, with much emphasis on the manufacture of capital equipment. In the past three decades, China has manufactured much new equipment unaided. In almost every industry, through self-reliance, plants large and small have been created. Special efforts have been made to spread manufac-

Table 9. Selected data for comparison

Item	Shanghai	Shaanxi	Gansu	Ningxia	Qinghai	Xinjian
GVIAO (1985)[a]	89.2	25.6	16.1	3.3	3.1	12.4
GVAO (1985)[a]	2.3	6.8	4.2	1.0	1.0	4.9
GVIAO (1985)[a]	86.9	18.8	11.9	2.3	2.1	7.5
Major sector output (enterprise unit numbers/ value output × 10^6 year)						
Mineral mining (all types)	4/4	704/777	492/980	121/387	61/68	457/124
Agro. food	754/3764	1599/1517	675/808	215/206	162/224	1243/1122
Textiles	844/14648	393/3082	120/556	46/135	20/152	309/1273
Handicrafts	155/816	163/49	117/51	20/6	36/19	110/26
Chemical engineering	433/5745	415/847	183/1110	64/101	29/96	101/182
Construction materials and non-mining manufactures	566/1768	1744/780	625/469	210/110	174/108	828/417
Metal processing	1034/11834	1004/957	608/2360	188/249	85/251	268/197
Machine-building	1586/11279	1178/2559	481/947	141/336	142/277	417/393
Electronics and communications manufacture	135/2357	6/132	4.100	1/9	–	3/38

a. In billions of yuan.

turing to backward regions. Meanwhile, industrial research institutes and the more advanced factories have striven to make new products and to master new technologies.

As a result, a solid foundation of engineering experience and a wide range of technical capabilities have been established. But owing to the rigidity of the economic and S&T system, advances have been made mainly through extensive growth. Technological self-reliance was sought not just at the national level, but also in individual ministries, provinces, localities, and even enterprises. The result of this was a wasteful allocation of scarce resources and slowness in product innovation and utility improvement.

The links between industry and agriculture were weakened in the past by the "grain first" policy. Cash crops were neglected for a relatively long period and this affected the development of agrofood and other agro-related industries. The linkage was unilateral in the past, industries providing the major inputs to agriculture, such as chemical fertilizers, diesel oil, farm machinery, etc., while the agricultural sector only provided primary products – the basic food for industrial workers.

There were 7,816,680 scientific and technical personnel at the end of 1985. The design capability of a country is an important measure of S&T capability. Science's impact on society is mediated through those professions concerned with design and construction. In 1981, there were 2,654 design institutes with more than 346,000 staff, and the construction industry's manpower had increased to more than 5 million. China had also established a large R&D force.

Theoretically, the country had already established a scientific and technological capability. Yet, industrial technology in China generally lagged behind that of the industrialized and newly industrializing countries. The current status of S&T and its rigid structure were derived from two sources: first, the Soviet model of S&T, inherited from the institutions and organizations that were set up in the first Five-Year Plan period, and, second, that derived from the political environment of the Cultural Revolution period. Modification of the Soviet-derived rigid structure took great effort and time, while the effect of the Cultural Revolution created an S&T gap with advanced countries, which also would take time to close.

Before 1949, there were just over 600 persons engaged in scientific research. Furthermore, research that was closely linked with production was practically non-existent. Yet, during more than 35 years of effort, through ups and downs, R&D has played an important role in

China's socio-economic development. R&D institutions were unable to realize their potential fully owing to the lack of horizontal linkages, the overcentralization of R&D forces in the Academy of Natural Sciences, the existence of different ministries, and the segregation of the civilian and defence sectors. Moreover, the traditional organization of R&D in China put too much emphasis on "technology-push" rather than on responding to demand. Consequently, exploitation of the technology market was proposed in the reform of the S&T management system.

There are six types of R&D institutions, as follows:

1. *The Chinese Academy of Sciences.* This has at present around 154 institutes under its organization.

2. *Ministerial and provincial research units.* These are scientific research organizations that function under various ministries of the State Council or those at the local level. For example, the Ministry of Water Resources and Electric Power has eight research institutes: the Electric Power Research Institute in Beijing; the Institute of Water Conservancy and Hydroelectric Power Research Institute in Beijing, which is jointly governed by the Academia Sinica and the Ministry; the Xian Thermal Power Engineering Research Institute; the Nanjing Automation Research Institute; the Electric Power Construction Research Institute; and the Nanjing Hydraulic Research Institute and Scientific and Technical Information Institute.

Another example is the railways. The Railway Ministry established the China Academy of Railway Science in 1950. This consists of 10 research institutes; Railway Transportation; Locomotive and Rolling Stock; Railway Engineering; Signalling and Communication; Metals and Chemistry; Computer Technology; Technical Information; Standards and Metrology; and South-west and North-west regions.

3. *The university sphere.* Research organizations in colleges and universities amount to around 1,400 units with 30,000 S&T personnel.

4. *The Factory Research Force.* These are research organizations run by factories and mines. Their main concerns are project-connected.

5. *Research Force in National Defence.* These are research organizations of national defence. The role of R&D in atomic research achievements, as well as in rockets and satellites, is well known.

6. *The Chinese Academy of Social Science (CASS).* This had 31 research institutes and 2,431 researchers in 1985.

In China, it should also be noted that there are very few private R&D institutions.

Table 10. Distribution of institutions and S&T personnel

	No. of institutions	Total no. of workers and staff	Personnel engaged in S&T activities		
			Total	Scientists and engineers	Other S&T personnel
Subordinate to departments under the State Council	622	266,412	204,370	93,026	36,787
Subordinate to provinces, autonomous regions, and municipalities directly under the central government	3,946	434,354	313,146	105,850	75,385
Subordinate to the Chinese Academy of Sciences	122	69,650	58,220	32,174	8,828
Total	4,690	770,416	575,736	231,050	121,000

Table 10 gives the distribution of S&T personnel and institutions under the various authorities.

Case-studies in the different economic sectors

Seven sectors in the economy were chosen for research in the economic sectors. The rationale for the choice of these sectors is given in figure 5. The case-studies focused on the agricultural sector, the light industry sector (textile and agrofood), the infrastructure sector (electric power), the heavy industry sector (iron and steel, machine-building, and machine tools) and the high-technology sector (electronics). Two illustrative case-studies are described in detail here, one on the agrofood sector and the other on the steel industry.

Food sector

Before its liberation in 1949, China did not have a fully fledged food industry. In the large cities there were only a few enterprises engaged in food processing, vegetable-oil pressing, livestock-slaughtering and cigarette-making, while in the rural areas there were only manual workshops. The country's food industry altogether numbered 1,379 enterprises with a total of 100,000 employees. Less than one-fourth of these enterprises operated with machine power and hired more than 500 workers.

For China, therefore, food processing is a new branch of industry. During the past three decades or so, the food industry has undergone three stages of development.

During the first stage (1949–1957), the industry received 3.7 per cent of the country's total industrial investment. Under the principle of cooperation between urban and rural areas and between industry and agriculture, many food-processing workshops were set up, both in the cities and in rural areas. During these eight years, the food industry grew at a rate of 13.2 per cent a year. In 1957, it accounted for 19.7 per cent of the country's total industrial output value, becoming the no. 1 industrial sector.

In the second stage (1958–1978), errors in the guidelines for national economic development – with their overemphasis on accumulation to the neglect of consumption, on heavy industry to the neglect of light industry, and on grain production to the neglect of cash crops and sidelines – resulted in a serious shortage of food raw materials, the slow development of the food industry, a reduction in food varieties,

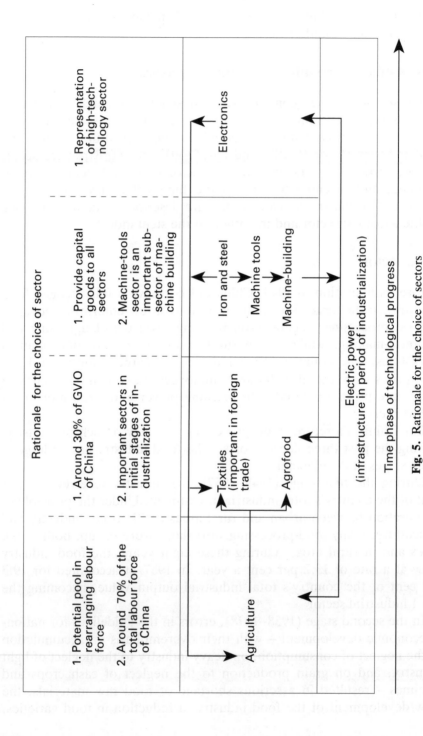

Fig. 5. Rationale for the choice of sectors

and a deterioration in quality. Food was now in short supply in the market-place. In this period, investment in the food industry was only 1.53 per cent of total industrial investment. In Shanghai alone, food factory buildings totalling a floor space of 400,000 square metres were used for other purposes.

In the third stage (1978 to the present), after the implementation of new rural economic policies, agricultural and sideline products have increased by a big margin and food raw materials have become unprecedentedly abundant. The food industry, to which the state now gives priority, has again taken the road of vigorous development. Investment in the industry constituted 2.96 per cent of the country's total industrial investment in 1980, 4.3 per cent in 1981, and 5.4 per cent in 1982, and continues to rise. The momentum of development in rural townships is even greater. From 1979 to 1982, the output value of the food industry increased by an annual average of 10.4 per cent. This was accompanied by an increase in food variety and a marked easing of food shortages in the market-place. By 1983, the food industry had diversified into 24 separate trades. There were more than 60,000 state-owned enterprises with 2.5 million employees, producing 6 billion yuan in output value. The food-processing sector led other industries in contributing taxes to the state. Food exports earned US\$ 3.5 billion.

Total output value registered an increase of 13.2 per cent during the first Five-Year Plan period (1953–1957); a decrease of 1.7 per cent during the second Five-Year Plan period (1958–1962); and increases of 11.4 per cent from 1963 to 1965, 2.4 per cent for the third Five-Year Plan period (1965–1969), 8.4 per cent for the fourth Five-Year Plan period (1970–1975), and 9.3 per cent for the 1979–1983 period.

Self-reliance in science and technology
The food industry suffered from various restrictions and developed slowly, and so cannot be expected to have made much technical headway. Technical progress in the field was negligible during the 20 years from 1958 to 1978. The present and future direction of the food industry is to achieve high-speed development on the basis of enlarged reproduction. The strategy aims to make up for long neglect by means of a sudden take-off. Therefore technical development is still directed towards popularizing existing techniques, with the import and development of new and sophisticated technologies taking second place. This trend is best illustrated by the proliferation of township food enterprises applying low-level techniques.

Though China's food industry has undergone unprecedented development in recent years, it is still very backward. The country's total output of farm products is in fact quite large, but the per capita amount is small and cannot be increased quickly. China is a success agriculturally. With less than 7 per cent of the world's cultivated land, it feeds nearly one-fourth of the global population. However, in comparison with that of other countries the speed of development is slow. Furthermore, China's average per unit area yield is lower than the world's highest, though it comes up to the world average. The per capita output of various kinds of food, particularly milk, is also lower than the world average.

At present, the output value of the food industry in developed countries is usually greater than that of agriculture. It is 169 per cent, for example, in the United States and 232 per cent in Japan. In China, the output value of the food industry is only one-third that of agriculture.

In the cooking process used by Chinese families, the discard rate is 20–30 per cent for vegetables, 30–40 per cent for fish, and 20 per cent for chicken and ducks. If the food industry is technically advanced enough, these leftovers can possibly be processed into food, or into feedstuffs, which can again be turned into food in the form of livestock or poultry. Because of a low capability in storage, processing, and transportation, a large amount of grain has to be stored in farmers' homes. This alone leads, because of insect pests, to a waste of 15 billion kg, or 8 per cent of the stored amount. China annually produces 8 million tons of fruits, but can store and keep only 5.3 per cent of them, resulting usually in a waste of 5–25 per cent. The wastage rate of vegetables is also 25 per cent. Similarly, the processing ability for aquatic products is less than 10 per cent of the output, causing an annual loss of 1 billion yuan. A lack of processing capacity for potatoes leaves 2.5–5 billion kg rotten, with a value of 0.5–1 billion yuan. The loss rate for grain products processed by traditional methods is 2 per cent greater than grain processed by machines. In 1980, however, grain processed by machines made up only 28.6 per cent of the total, and as a result some 5 billion kg of grain was wasted. Compared with the traditional pressing technique in edible oil extraction, the soaking method gives a 5 per cent increase in output, but only 13 per cent of oil-bearing seeds are now processed by the soaking method, leading to a loss of 350 million kg of vegetable oil.

This inadequate production capability is closely connected with the low level of investment in the industry. The food industry in Shanghai, for example, has given 9.8 billion yuan to the state over the past 30 years, compared with only 50 million yuan invested by the state in food enterprises during the same period. As a result of the lack of investment, food factories have to operate with out-of-date equipment, which often breaks down. At present, with the situation in rural areas getting much better and offering greatly increased amounts of farm products, shortfalls have been replaced by overabundance. The consequent stockpiling of farm products calls for an urgent, large-scale development of the food industry.

There is no difficulty in developing suitable technologies in the fields of storage, transportation, and primary processing. What is needed is investment funds and the popularization of technologies. Because the accumulation of funds needs time, low-level and even primitive techniques will continue to exist for some time yet.

A surplus of certain farming products cannot by itself meet the demands of the food industry. A large-scale food industry needs large-scale, regional raw materials and a supply base that is guaranteed by the cultivation of suitable varieties. In other countries, food enterprises usually own their own farms or have long-term contracts with farmers. The enterprises decide what to plant, how much, and when, and even dictate the kind of fertilizer to be used and the time of harvest. For example, in the United States, the state of Florida is a producer of oranges, the state of California of peaches, and the state of Hawaii of pineapples; in Japan, the counties of Ehimi and Shizona yield great amounts of tangerines, while Hokkaido is a producer of asparagus. Such specialized economic zones make it possible to set up large-scale, highly efficient and highly mechanized canned-food factories, wineries, and sugar-making factories.

China lacks adequate raw material bases. Structurally, the food industry is characterized by scattered raw material supply bases and small-scale operations. In terms of economic relationships, the practice in China is the direct opposite of that in other countries: here it is the farms or farmers that run food enterprises, instead of vice versa. The existing supply bases, like the green bean base near Nanjing, now face serious difficulties. Many food factories, set up on the basis of "do whatever is available," simply cannot expand in scale and become highly efficient.

The comprehensive use of raw materials, in effect, increases their quantity. Technical treatment immediately after the harvest and timely industrial processing can prevent the decay of grain and reduce losses. Modern processing technologies can turn inedible ingredients of grains into edibles or materials of higher value. Technologies for the comprehensive use of raw materials are as yet far from being developed in

China. Neither is enough attention paid to the resources of waste products. Waste matter and left-overs from the country's gourmet powder factories, sugar-refining factories, butcheries, and breweries alone are enough to feed 4 billion chicken each year.

Some countries in the world have developed unique food-processing industries for export based on imported raw materials combined with advanced technologies. Denmark's fruit jams, the Netherlands' cocoa and Switzerland's instant coffee, for example, are all products made from imported raw materials. China has yet to build such industries.

In order to create an industrialized food industry and raise economic efficiency, large-scale raw material bases must be set up in China, and varieties of food crops suitable for industrialized processing cultivated. To demonstrate the distance yet to be covered, we could cite the example of potatoes. Potatoes processed for their starch in some countries contain 22–24 per cent of starch, and some new varieties reach a figure as high as 28 per cent. In China, however, the starch content is only 12–20 per cent.

At this stage in the development of the country's food technology, China has engaged in a series of programmes including directional breeding, technical guidance and services, the popularization of fine crop strains, pricing policy, long-term purchase contracts, and regional planning.

Since the shortage of food resources is an important factor limiting the development of China's food industry, continuing neglect of the establishment of raw material supply bases will only perpetuate this shortage. This in turn will restrict the development of a modern food industry. On the other hand, with advanced technologies we can enlarge the sources of food raw materials through comprehensive utilization, high-tech processing, and the development of a sector which processes raw materials for export.

Except for the consumption of food grains, which is higher in China than in other countries, the per capita consumption of other foods is quite low. And except for sugar, which has a 90 per cent commodity rate, the rate for market-oriented foods in China is generally low: 14.25 per cent for grains, 43.7 per cent for pigs, 1.95 per cent for beef cattle, 6.1 per cent for mutton, and 53.1 per cent for cooking oil.

The present low level of consumption offers great potential for the development of the food industry. Food science and technology, therefore, have considerable market prospects. For instance, there is stockpiling of such starch-containing products as maize, potatoes, and cassava in one part of the country, while elsewhere starch, noodles, and vermicelli are in short supply. This can be avoided through proper utilization of the industry. The extremely low level of commercialized food products demonstrates that the food industry is facing challenges from a natural economy and from traditional consumption patterns. A solution to this problem depends on sustained economic development and greatly improved living standards.

China's food industry is not complete. This is reflected in a weakness in the supply of basic raw and accessory materials for the flour, oil, protein, sugar, and food additive industries.

The flour industry in the United States turns out about 100 varieties of flour. In Japan, 64 varieties of flour are used for making bread. China has only two; a coarse type and a fine type. In cooking oil, there are dozens of varieties, and these, hydrogenized to varying degrees, can be mixed to make hundreds of varieties. In Japan, for example, cooking oils number more than 200. In China, however, there are only two types; unprocessed raw oil, which dominates the market, and a small amount of second-grade oil.

There is as yet no protein industry in China. Every year, the 2 million tons of cottonseed cakes and 1 million tons of rapeseed cakes, as well as peanut and sesame cakes – all by-products of oil pressing – are mostly used as animal feed and manure.

Maize, a source of sugar after sugar-cane and beet, is a staple crop in China. Glucose, maltose, levulose, top-levoluse syrup and other raw materials needed by the food industry can be produced from maize starch. But China has not fully utilized maize. The starch industry remains at a simple level and is still confined to turning out one product – raw starch. Starch reprocessing is very limited and no development has been made toward making denatured starch and starch derivatives. Of the starch used overseas, only 30 per cent is raw starch, and 70 per cent is reprocessed starch. The number of starch derivatives exceeds 100.

Food additives including trace elements, microbial elements, pigments, germicides, thickeners, emulsifiers, and special condiments are very important in raising food quality. They are widely used overseas. In China, there are fewer varieties and their output is small. "Grandma Bean Curd," a popular dish from Sichuan province, has become one of the main soft-can foods in Japan simply because a special condiment with the flavour of "Grandma Bean Curd" has been developed and mass-produced there.

In short, because of the rough and simple processing of farm products in China, food raw materials are not fully and rationally utilized.

In the meantime, it is difficult to turn out high-increment products owing to the shortage of necessary food raw materials – particularly of special basic ingredients – that results from an absence of intensive processing.

The food industry in the United States developed 9,747 new products between 1970 and 1979. In Japan, there are more than 300 kinds of bread and over 1,400 types of canned food. Products on foreign markets include cholesterol-free, sugar-free, and salt-free foods, low-caloric food, and low-sodium food, as well as high-energy, high-protein, and high-cellulose foods. There is also food for the young, the aged, and athletes, and foods with various kinds of local flavours.

Although an increasing number of new food products have come into being in China in recent years, those foods sold in volume have remained the same as before. Baby food and fast food have not as yet established a reputation. Many new products often disappear shortly after their introduction.

Decades of neglect in the food industry have resulted in a huge gap between supply and demand; and the development of basic raw and accessory materials and new products is largely out of the question. Coordination among specialized enterprises does not yet exist in China, owing to defects in the economic structure as well as to China's traditions in the food-processing sector. A large number of newly established enterprises are gearing their production directly to the consumer market, while the basic materials industry attracts little attention. As a result, the limited number of food scientists and technicians have not been able to devote themselves to strengthening the foundations of the food industry.

Food equipment manufacturing is a backwater of China's industry, receiving little attention and developing at a slow pace. Except for the canned-food, dairy, soft drinks, grain-processing and oil-pressing industries, whose equipment is partially provided by specialized factories, the old factories in other fields have to depend on their own repair and spare-parts workshops for new equipment. It was not until 1978 that industries began studying the problem and started manufacturing food-related equipment.

The present situation can be described as one of a weak foundation, low technological standards, and a lack of sophistication in equipment. The development of equipment that carries out intensive processing and permits a comprehensive utilization of raw materials has only just started. China's food-equipment manufacturing industry is backward. Canned food production in China, for example, is a sector with a relatively high level of mechanization: each worker produces 4.7 tons of canned food a year. But this is only one-forty-sixth of what a worker in the United States produces. The efficiency of Chinese machines for making bottles is only one-tenth that of those made overseas. China-made machines can pack only one-quarter the amount of candies packed by similar machines.

China is capable of making small renovations to simple separate machines or production lines and gradually copying them. In 1980 the Baiyun Rice and Flour Products Factory in Guangzhou introduced a fast-noodle production line. Within a year, the factory was able to pay off the foreign investment of US$280,000. The factory renovated the "irrational" parts of the production line, raising its efficiency as well as the quality of fried noodles. Moreover, all the spare parts that are susceptible to wear and tear can be substituted by those made in China. In 1983, China's first fast-noodle production line was successfully trial-produced, passed appraisal, and was put into operation in Tianjing. The production line cost only 270,000 yuan. A renovation was made on the production line later, changing de-watering through frying into de-watering through steaming. The change gave birth to a steamed fast-noodle production line. By July 1985, 60 steamed fast-noodle production lines made by the Renmin Machinery Factory of Guangzhou had been installed and put into operation in many cities. In the food industry, there are a good many facilities of this kind which can be manufactured by unspecialized factories. However, despite the fact that China has the ability to do this, repeated imports of the same technology are still a common practice. Up until now, about 40 fast-noodle production lines have been bought. By the end of 1983, foreign investment used by China's food industry for introducing advanced technologies amounted to over US$100 million.

The New Sugar Refinery in Jilin province was a major project in the first Five-Year Plan period (1953–1958). All the facilities and technology were supplied by Poland. The refinery was designed to treat 1,000 tons of beet per day. Technical renovations and the addition of equipment increased the refinery's capacity to 1,500 tons by 1980. The technical level and economic efficiency of the refinery have always been counted as high in China, yet its equipment and technology, in fact, belong to 1940s Poland. In 1983, the refinery introduced from Denmark a cleaning technology, seven key pieces of sugar-refining equipment, and six control systems. These facilities, after some technical transformations, can treat 3,000 tons of beet per day.

The equipment in the above example is typical of that currently used in the food industry. Thirty years ago, China built part of its food industry with imported equipment. Factories operating with such equipment have given the best performance during these 30 years. Now China's food industry is 30 to 40 years behind the developed countries, and it therefore has to continue importing advanced equipment to catch up.

Food factories can make small renovations to existing machines, and specialized equipment manufacturing factories have the ability to copy and improve simple machines, but they are unable to manufacture highly efficient, technically advanced equipment in complete sets, let alone develop new models of machines. This is because the latter are closely related to the general level of the technology employed by the food industry and the scale of that industry. There will be no demand for highly efficient equipment without the existence of large-scale enterprises.

Packaging is an indispensable part of high-grade, high-increment products. Food and food-packaging industries promote each other. China still has to import many packing materials and containers. To some extent, foreign countries can even influence our food production and export by controlling the export of packaging materials to China. China's export foods still face the problem of "first-rate materials, second-rate quality, third-rate packaging, and fourth-rate price," giving poor economic results.

With regard to glass containers, the trend overseas is toward lighter and thinner types strengthened by chemical processes. Improved sealing technology has given rise to various kinds of push-turn covers and plastic-metal composite caps. We still use old-type caps that are difficult to open, in spite of repeated complaints from consumers.

Overseas tin-plate making has developed from double cold rolling to continuous casting, and from electroplating to organic polyester coating. China imports large amounts of material for tin plate. Can-making overseas has developed from double reeling to deep punching, three-segment connecting, and welding. These technologies are still not used in China, and only a few sample machines have been imported. Owing to an inability to control lead penetration in the welding process, Chinese-made cans face the danger of being squeezed out of the international market.

The development of compound material for plastic packaging takes place even more quickly overseas. The appearance of plastic tins is due to the invention of compound materials able to withstand high temperatures. Different layers of this material play complementary functions, resulting in an ideal packaging material for food. In China, the quality of the paper, aluminium foil, and plastic used for compound packing materials is not yet up to standard. The so-called "soft packaging," with only one layer of plastic, that is seen on the market does not meet hygienic requirements.

In general, the cost of packaging does not exceed 10 per cent of the total cost of production in foreign countries; the cost for ordinary food varieties may come to only to 3–4 per cent of the total. The main way to cut down on packaging costs is to set up highly centralized packaging materials and containers enterprises. Some such enterprises overseas have developed into multinational corporations. Packaging materials in China, owing to price hikes for raw materials, can account for 30–40 per cent of the total production cost of food, adding to the burden on consumers and hindering the development of packed food.

The more intensively food is processed, the more coordination among specialized departments is needed. This is one of the weakest links in China's managerial system and economic structure. Food technology is subject to packaging technology, which in turn is subject to the materials industry.

Some of China's food products are comparable with, or even better than, those overseas. The quality gap in the final product is often due not to technology per se, but to raw and accessory materials, packaging, equipment, factory conditions, and circulation links.

It must be admitted, however, that the technology in most factories is backward. This reflects a weakness in the development of technology which is based upon experience rather than a systematic study of mechanisms. Consequently, some operational processes cannot be flexibly adapted or modified to guarantee the quality of products. This indicates an inadequacy in technological personnel, which hampers the popularization of more mature technologies.

In our push for the industrialization of traditional foods, the research and development of technology is also crucial. Technology should not be just a means of expanding traditional methods; it involves different, modern methods which can nevertheless be based on the mechanisms of the old. China's traditional method of making bean curd is worth studying. It is now not as advanced as those available overseas because of a shortage of technical personnel and research means.

The starch industry demonstrates the backwardness of the technology. In foreign countries, a closed process is used, which needs less water, discharges less waste water, employs technologically advanced equipment, and is highly efficient. The advantages of the new production method are a high rate of starch collection, high quality, and a better comprehensive utilization. The utilization rate of maize reaches 99 per cent. However, in China, the open process is still used. It needs large quantities of water, discharges too much waste, causes pollution, has a low rate of starch collection, and produces starch of low quality. Additionally, many exploitable resources are wasted in those small factories with a low utilization rate. Production of 500 kg of starch from potatoes, for example, involves a waste of 80 kg of protein.

In other countries, raw materials are processed at multiple levels, which results in their complete utilization. For instance, in processing kelp Japan makes some 60 different kinds of pickles. Even the water used for boiling kelp is made into a soft-packed food. In China kelp is merely dried in the sun before being marketed, which involves no technology at all.

The existing technology is not widely applied, and this sometimes causes big technological gaps between factories in China. This phenomenon, not limited to the food industry, is due to an economy of shortages and a lack of competition caused by underdeveloped commodity production. As a result, technically backward enterprises have the "right of existence" and similar factories still continue to be built. There is not enough motivation to raise comprehensive economic efficiency, nor is there pressure to do so.

Up to now, China has not had a short- or long-term plan for training technical personnel for its food industry. A very limited number of students are being trained in the country's universities and colleges to become technicians and food industry administrators. In the light industrial sector, which has the biggest number of trainees, only 2,682 persons were trained in institutions of higher education during the sixth Five-Year Plan period. During the same period, about the same number were trained in secondary technical schools. If the situation remains unchanged, there will be only one technician for every two food factories by the end of this century.

In the 60,000 state-owned food enterprises, technical personnel make up less than 1 per cent of all staff and workers. Technicians with a college education are very scarce. The 300,000 collectively owned enterprises have their own comprehensive food research institutes. Three provinces and autonomous regions have no food research institutes of any kind. Most of the existing research organizations deal with the primary processing of grain and edible oil.

Take the starch industry, for example. Many countries have set up national research institutes. There is the Northern Agriculture Research Center in the United States and the National Grain and Potato-processing Research Centre in Germany. Companies or factories dealing with starch or starch sugar, as well as some universities, have set up their own research institutes. There is not a single research body in China's starch industry, nor do universities and colleges offer courses related to starch production.

We can also make comparisons with Japan. From central to local departments and to enterprises, Japan's food industry has an established system of research institutes. Located in the science city, Tsukuba, is the state-run general food research institute, which operates under the jurisdiction of the Ministry of Agriculture, Forestry, and Aquatic Products Industry and has about 100 senior researchers. Every county has a food research institute with 10–40 researchers. Most of the food enterprises have their own research and development department with researchers accounting for 20–40 per cent of the total workforce. All the research institutes have a rational division of work and cooperate with one another, avoiding repetition and waste. The research budget of Japan's scientific research food industry in 1980 reached about 1 billion Renminbi (RMB). Japan also has a complete educational system to train qualified personnel for the food industry. Many universities have courses in food disciplines, while some big enterprises have set up training schools and food colleges offering two-year courses.

The lack of scientific and technological personnel is normally a major factor affecting self-reliance in science and technology. This problem is more serious, however, in China's neglected food industry. A way out lies in reforming the economic structure to allow researchers and technicians in the food industry to develop their talent to the full by occupying proper posts. China's old administrative system makes impossible a rational distribution of technical and research personnel. The latter face all sorts of possible technical problems in individual factories. Meanwhile, the technical personnel in research institutes often have to study basic problems in a state of isolation. This results in a failure to pool research personnel for major projects, which could have offered a way of solving the problem of the general shortage of personnel.

Further development of food science and technology, and of the food industry itself, cannot be assured if necessary measures are not taken in macro-management. What is needed is a feasible programme for the development of the food industry, including programmes for different trade lines and a science and technology development plan. At present, food production is scattered throughout a dozen or so economic sectors including light industry, commerce, agriculture, animal husbandry, fishery, and land reclamation. As a result, it is even more difficult to make a coordinated and feasible general plan for the industry.

Second, there must be a definitive policy for the industry. Whether it is categorized as being closer to agriculture – providing basic food for the people – or to industry – offering high-grade consumer goods – will determine its scope. Overseas, canned food is a low- or medium-grade consumer good, while in China it is regarded as high-grade. Many agricultural products are developed because of price subsidies. The food industry often has to use raw materials at a negotiated price because of a shortage of allocated, low-priced materials. Price gaps put factories producing industrial foods at a disadvantage, compared with those producing primary agricultural foods. High tax rates have hindered the development of some food products. For instance, the tax rate for maltose and glucose is as high as 30 per cent.

Coordination among specialized enterprises is essential to the development of large-scale socialized production. Coordination among various fields, and links between production and trade, are especially needed in the food industry, particularly by enterprises doing intensive processing. An absence of socialized coordination necessitates the maintenance of a workshop production method. China's economic mode, structure, and management system have forced enterprises to try to do everything themselves, regardless of their scale. The food industry is no exception. Overseas, canned food factories do not produce their own cans. In Japan, for example, all empty cans are provided by three specialized factories. All the cans have brand names printed or pasted on, so that the canned-food factories have only to do the packing. A production line making a single type of can helps standardize the product and makes it easier to employ advanced automation facilities. In China, however, all the 700 canned food enterprises make their own containers.

There is an even larger gap between China and developed countries in micro-management. In Japan, one yen of fixed assets in the food industry can give an annual output value of 35 yen, while in China one yuan generates only 3.7 yuan, one-tenth of Japan's output.

Iron and steel

Iron and steel is a traditional industry. China had already mastered the problems of production, construction, and design capability. The present problem is insufficient production capability, leading to an increase in the importation of steel in recent years. For example, steel imports were 12,300,000 tons in 1984 and 20,030,000 tons in 1985. To illustrate the problems in the industry, a micro case-study of the Capital Iron and Steel Corporation, an enterprise following the overall economic responsibility system, is presented.

The Beijing-based Capital Iron and Steel Corporation (CISC) is an iron and steel complex whose activities include mining and iron-ore dressing. CISC's predecessor was the Shijingshan Ironworks, which was established in 1919 and produced only iron before 1949. The aggregate output of iron for 30 years was a mere 286,200 tons. The plant was rebuilt after liberation. Now CISC mines and dresses 14 million tons of iron ore, turns out three million tons of iron and 2.7 million tons of steel, and rolls half a million tons of steel. The corporation now markets more than 200 brands of steel products. There are 14 categories and more than 100 varieties. CISC made a profit of 430 million yuan in 1983.

The overall economic responsibility system for the enterprise has two major components. First, the enterprise promises, by contract, to fulfil the tasks and targets stipulated by the state on the basis of increasing its profit turnover to the state by an annual rate of 6 per cent, while the state gives a corresponding degree of autonomy, as well as economic benefits, to the enterprise. Second, under the responsibility system adopted by the corporation, responsibilities are spelt out for executives of the corporation, plants, and mines, as well as for workers in workshops and grass-roots units. Their work performance is strictly vetted, and rewards and punishments are handed out.

The economic reforms have brought changes in both the internal and external conditions of the enterprise's technical work. Under the old system, most of the profits and depreciation fees had to be turned over to the state. No funds were available for technological development, technical items had to be approved by the authorities concerned, and any technological progress made few gains for the enterprise. Now, CISC returns a greater amount of profits, and uses its own money to raise the technological level of its plants.

Within the corporation, a technical responsibility system links the interests of managers, technicians, and employees to technical progress in their respective fields. This motivates all to engage in technical innovation and to achieve good results. Practice has proved that the economic reform has promoted and accelerated the technological progress of the enterprise. From 1978 to 1984, CISC, using its own capital, achieved good results by adopting a set of new technologies in the main production system. In 1984, of the 70 major comparable indexes in terms of technical level and economic efficiency in the metallurgical industry, CISC was among the top 35. In iron-smelting and converting, some technologies have reached an international level. The profits of the enterprise have grown by 20 per cent annually for six consecutive years.

Technical management is primarily the responsibility of the corporation's Technical Department. Its main task is to make technology development plans, organize and coordinate the carrying out of the plans, and conduct the day-to-day management of technical work. The department is under the direct supervision of the deputy manager of production and technology and the chief engineer. Specific research and design projects are initiated by the Design Institute and the Research Department for Computers, Control Equipment, and Instruments. There are corresponding technical sections and technicians in plants, mines, workshops, teams and groups. Thus, a four-level technical management system is created, from the Technical Department down to the work groups.

Technical management work consists mainly of implementing technical planning, technical-specialized planning, technical quotas, regulations and standards, technical analysis, and technical measurements.

Technical planning is an important integral part of the corporation's production development planning, and oversees technical principles, quotas, measurement, trial production of new products, and scientific research. It is also concerned with reducing energy consumption, improving quality, developing scientific information, and popularizing science and technology. An annual and a long-term plan is worked out.

On the basis of this technical planning, a technical-specialized plan is made by the Technical Department from initial plans sent by plants, institutes, and centres in order to meet the requirements of scientific research and to exploit new products. The technical department controls technical quotas, measurements and standards.

The technical analysis keeps abreast of developments in production and technology, determining what affects output and quality. Technical renovation upgrades old enterprises, with a consequent increase in benefits. Until the end of 1978, the development of CISC took place mainly through new construction; after that, the focus of its technical work was on technical renovation. The enterprises had limited funds and old equipment on which fresh investments would have little effect. CISC selected, designed, and evaluated items for renovation, taking several measures designed to tap the potential of the enterprise and to increase output, develop new products, improve quality, lower energy consumption, find new ways of saving energy – such as tackling waste gas, waste water, and industrial residue in a comprehensive way – and protect the environment. Some concrete examples of these improvements are given below.

Tapping potential to increase production
The steel plant originally had three 30-ton converters, with a designed capacity of 600,000 tons. By using new types of furnace lining, enlarging the furnace volume, prolonging the furnace lifetime, and applying composite blowing, the plant reduced consumption and shortened steel-making time. Consequently, annual output in 1984 climbed to 1.67 million tons. The 1985 production was expected to reach 1.8 million tons, 300 per cent of the designed capacity.

Developing new products and improving quality
Measures were taken to improve the quality in every aspect of mining, sintering, steel-making, and steel rolling. Through the application of a computer-controlled production process, the concentrate grade was raised from 62 to 68.3 per cent, the qualified rate of sintering jumped from 76.6 to 97.5 per cent, and the left-over rate for cutting steel billet was lowered from 3–3.5 per cent to 1–1.5 per cent. In 1984 alone, more than 140 new products with 418 kinds of specifications were developed.

Reducing energy consumption
The iron and steel industry accounts for about 13 per cent of the total energy consumption of the national economy. In the iron and steel complexes, the process of steel-making takes 45 per cent of the total energy consumption; with sintering and coking, it amounts to about 70 per cent. So the Capital Steel Plant centred its energy-saving efforts on the iron system, first attempting to lower the coke ratio. Using new techniques for emitting coal powder, the furnace coking ratio was cut from 455 to 412 kg, with an annual profit of 16 million yuan. This was

only one-third of the investment. After the renovation of the furnace and the heating furnace, which use converter-coking coal-gas mixed supply steelrolling, the diffusion rate of gas produced by four converters declined from 17.9 to 8 per cent, the utilization rate was raised, and pollution decreased. The pressure difference was also used to generate power and hot water for heating purposes.

Environmental protection and comprehensive treatment of waste gas, waste water, and industrial residue
Environmental protection is a key aspect of technical innovation in the complex. In fact, environmental protection also saves energy, increasing production and improving work conditions. For instance, the technical renovation of the sintering machine of No.2 blast furnace is just such an item with additional benefits. In recent years, the Capital Steel Plant has focused its work on raising the water recycling rate and reducing the waste water discharge and dust concentration in gases. More than ten synthetical treatment programmes were carried out, all with obvious effect. The phenol contained in the waste water discharge declined from 8–10 mg/l to 0.04–0.06 mg/l. In the sintering workshop, where dust smoke is a serious hazard, the dust concentration in the gases declined from 1,000 mg/cubic metre to 100–150 mg/cubic metre; the dust concentration of gases in 90 per cent of the workshops is less than 10 mg/cubic metre, up to international standards.

As an major enterprise in a developing country, the Capital Steel Plant should aim to learn from the most advanced enterprises in the world, relying on its own abilities and improving itself step by step. By adhering to this principle in renovating its No.2 converter, the plant achieved the comprehensive objective of high output, low energy consumption, environmental protection, and automation of production.

The old No.2 blast furnace of the Capital Steel Plant was made in Japan in 1929. It was moved to Beijing from Pusan, Korea, in 1941. The blast furnace body is of the steel belt type, with a single-bell furnace crown and horizontal raw material feeding. It was still very old-fashioned after several major repairs. The furnace coke ratio was as high as 600 kg, combined fuel was 700 kg, the coefficient of use was 1.4 tons per cubic metre per day, and the pollution was heavy. In 1978, the plant decided to carry out major repairs on the converter. Considering the time that the repair would take – eight to ten years – and the fact that China was at least ten years behind the world level in converter technology, it was decided instead to perform a comprehensive technical innovation.

The innovation began in 1978 and was completed by 1979. The renovated No.2 converter adopted 37 items of new technology, including a blast furnace with bell-less top, injection of coal dust as fuel into the blast furnace and crown-burning hot gas furnace, and other environmental protection measures. The furnace volume was enlarged from 500 to 1,327 cubic metres. In 1984, an imported programmed controller and other computer systems were installed. The material supply, coal injection, and the main body of the blast furnace were automatically controlled. The renovation proved to be effective. The technical and economic targets approached international levels. The total investment was recouped within 19 months.

An overall survey of the Capital Steel Plant had shown that the plant lagged far behind advanced international levels in respect of the technical equipment in every system. The mastering of advanced foreign technology, therefore, was always a key task for the plant.

The technical information network is the chief organization for obtaining foreign technical information. A chief engineer is in charge of the network, which is run by the technology department. The designs institute of the information office is the backbone of this information network. Subordinate plants and sections are responsible for collecting information in their own speciality. Every information unit puts forward a comprehensive information summary data to the technical department every six months. This serves as a basis for scheduling technical design and keeping up with, and surpassing, advanced levels both at home and abroad. An annual session of the information network is held every October to exchange experiences and discuss the plan for the following year. The network solves important scientific and technological problems faced by the enterprises.

The "importation office," which supervises the import of technology, is also an important information service. The network attends to purely technical information, while the "importation office" deals with the analysis of prices and commercial information. Consultative arrangements with foreign companies and experts, overseas investigations, and attendance at conferences and exhibitions are also channels for obtaining information.

The direct import of foreign techniques is an effective way to improve the technical level of an enterprise. The process of technical importation includes collecting and analysing information, choosing and evaluating techniques, negotiation on import items, importing, installing, and testing. In this way the managerial and technical ability of the enterprise is improved.

Between 1981 and May 1985, the Capital Steel Plant imported 122 technical items, of which 15 brought obvious benefits; the volume of business amounted to US$100 million. To ensure correct technical importation and digestion, the Capital Steel Plant set up an "importation office," which organized the import of technology, and collected, analysed, and sorted the information. It was ensured that the technology was up-to-date and appropriate for the plant. Stress was laid on the importation of instruments, control systems, and equipment that would rapidly improve the technical level and reasonable use of resources. Investigations were conducted to compare different technologies before importation so that both advanced technology and a reasonable price was guaranteed. Attention was paid first to absorbing and then to transferring the imported technology.

Technical innovation in old enterprises is closely related to the dissemination of technology. The lack of knowledge of new technology is an important factor that delays its spread. Awareness of existing technical systems is a prerequisite for technical innovation. A major factor influencing the effectiveness of technical work is its organization and systems. In this lies the significance of the economic reform and economic responsibility in the Capital Steel Plant.

The analysis of the Capital Steel Plant shows that self-reliance in technology means relying on the development of the state's, enterprise's, or scientific research unit's own resources – that is, their knowledge, equipment, organization, personnel, and investment. Self-reliance results from the combined effects of organization, absorption of information, technology development, policy-making, and policy implementation. It is not an individual ability, but a social one.

Exogenous sources for technological progress and self-reliance

As members of modern society we live in an interdependent world. Owing to the rapid development of transportation and communications, distances have become shorter and the degree of interdependence has increased. There are interdependencies between regions and nations as well as between provinces and regions within a country.

It is not easy to identify the feasibility for S&T interdependence in specific areas and sectors. Only general observations can be made.

The sectors highlighted in the case-studies were: agriculture; light industry (textiles and agrofood); infrastructure (electric power); heavy industry (iron and steel, machine-building and machine tools); and high technology (electronics). Economically, China has a static comparative advantage in the agriculture, agrofood, and textile sectors. It is therefore possible in these sectors to cooperate well with exogenous sources in order to achieve technological progress.

A full range of production, R&D, and design capability has been built up in nearly every sector of economic activity in China. It is now necessary to strengthen the horizontal linkages and to emphasize the interdependence between sectors and activity components for mutual technological progress. Owing to the prolonged closed-door policy in the past, industrial technology in China has generally lagged behind that of industrialized and newly industrializing countries. Even in the traditional agrofood sector, it is still necessary to draw on exogenous sources in the drive for technological progress.

Cooperation with the outside world does not mean abandoning self-reliance. Interdependence and self-reliance are complementary; self-reliance should still be emphasized, because in terms of S&T the developed countries have a comparative advantage. Many sophisticated technologies are part and parcel of the military R&D of developed nations, and so governments impose controls and restrictions on them, mandating an emphasis on self-reliance among less developed nations. Developed countries, for their part, reap the benefit of "job creation" in their economies when they engage in technology transfer.

Past technology transfer can be roughly divided into three periods.

First, in the 1950s, China adopted a "follower's" strategy, importing technology from the Soviet Union and other Eastern European countries. This transfer was in the form of both hardware and software, and complete sets of equipment and processes were imported. Emphasis was laid on the capital goods production sector. Technology transfer also took the form of complete sets of technical drawings and documents and the services of expert technical assistance in planning, design, construction, testing, and the operation and maintenance of various types of industrial plants. The establishment of research organizations was also assisted by foreign experts.

Second, in the 1960s, after the Sino-Soviet conflict had come into the open, China imported technology and equipment on a small scale from Japan and the Western European countries. Priorities were placed on metallurgy, chemical fibre, petroleum, chemical engineering, textile machinery, mining equipment, electronics, and precision machines. This importation of technology supplemented the technological capability already established in the 1950s, and promoted S&T self-reliance in textiles, machine-building and machine tools, and electronics. Self-reliance and the process of "walking on two legs" were emphasized during this period.

In the period 1956–1967, a 12-year S&T development programme was carried out. This provided China with a technological capability in nuclear energy, jet technology, computers, automation, semiconductors, and radio electronics. The success of the programme was due not only to the coordination of different sectors, but also to the coordination between different regions, particularly with the relatively developed regions such as Shanghai. In this period widespread development of small- and medium-scale labour-intensive industries in the countryside was also initiated.

Third, in the 1970s, after she had had her United Nations seat restored in 1971, China imported technology and complete plants from a variety of non-communist sources and on a larger scale than ever before. This had a strong positive impact on the productive capability of several industrial sectors, such as chemical fertilizers and chemical fibres. But the absorption of this technology remained a problem, owing to the rigidity of the existing S&T system.

Since the announcement of the basic policy of "Opening to the Outside," China has started transferring technology from abroad on a larger scale, adopting various means and forms. Steps such as the encouragement of foreign investment (such as joint ventures, cooperative ventures, direct investment, and leasing) and the opening of Special Economic Zones (SEZ), including the 14 cities and the three deltas along the coastal area, all involved certain amounts of technology transfer. Whether China can benefit further from this policy or not depends on the establishment of a proper system of related policies, which will be mentioned later.

The effort to achieve technological progress based on exogenous sources and "self-reliance" is best described in the following quotation from a Western expert:

While the realization of the concept of "walking on two legs" has differed in degree from phase to phase and according to the political line prevalent at the moment, it has never been subject to any real basic doubt. Traditional, labour-intensive methods were employed in every area where they could make an effective contribution to development; . . . on the other hand, China did not seal itself off completely from large world-wide technological developments. China was aware that its level of performance in certain technological

Table 11. Innovation by source (international experience) (percentages)

Country/region	Year	Innovation source	
		Government and university	Enterprise
USA	1953–73	5	80
Europe	1982 (report)		70

fields was limited and therefore made some effort, in fields important to development, to catch up with international standards.

The proper selection of the right type of technology is the key to upgrading technology. In the past, China had swung between the extremes of importing turnkey plants and relying wholly on domestic R&D. And the traditional organization of R&D in China involved a top-down technology push. This approach was successful in limited cases, but international experience has proved that individual enterprises play a much more active role in innovation, as shown in table 11.

This is also demonstrated by the case-study example of the Capital Iron and Steel Corporation, which was one of the first groups to experiment with the new economic system. It had more autonomous decision-making power than enterprises had previously, and S&T played an increasing role in the promotion of its production and profits. Clearly, there is a need to push the economic reforms further.

The segregation of sectors in the past has seriously affected the utilization of exogenous sources in a broader sense. Domestic sectors and regions were isolated from each other, as was the defence sector from the civil sector. Major advantages can be gained by breaking down these barriers.

Central government agencies and ministries, with their research institutes and superior resources, have played a dominant role in determining what innovations are needed. This top-down "technology-push" approach to innovations was effective in large programme efforts, such as those in the period 1956–1967. But it failed to meet the needs of a wide variety of commodities. Large enterprises such as the Capital Iron and Steel Corporation can solve such problems alone. But for the industrial system as a whole, responding to the demands of users is important in promoting the self-reliance capability of R&D. This is the idea behind developing a technology market, as suggested in the S&T management system reform.

125

A desirable path and a strategy for S&T development

China has announced the "Decision of the Central Committee of the Communist Party of China on the Reform of the Science and Technology Management System," which is an official document outlining the new strategy for S&T development. One should mention here that the S&T system is a subsystem of the economic system. The economic reform should provide a suitable environment for the current S&T development strategy.

The aim of the economic reform is to establish a dynamic socialist economic structure (including the establishment of an organizational structure for the S&T system). The planning system (including the S&T planning system) will be reformed so that the law of value is consciously applied in developing a socialist commodity economy.

In the economic sphere, special emphasis is placed on invigorating enterprises and establishing various forms of economic responsibility. The parallel in S&T is the reform of its funding system. The reform of the economic management system includes separation of the functions of government and enterprise, promotion of a new generation of cadres, and the continual expansion of foreign and domestic economic and technological exchanges. China will no longer isolate herself from the economic and technological world. It has been realized that the S&T system of the industrialized countries evolved organically with the rest of their socio-economic systems.

The current strategy for the S&T system, outlined in the "Decision on the Reform of the S&T Management System," contains two basic elements. First, the reform is based upon general guidelines which specify that "science and technology must serve economic development, economic development must rely on science and technology." The three main aspects of this are reforms of the operating mechanism, organizational structure, and the S&T personnel system.

Second, the reform of the operating mechanism of the S&T system entails the reform of the funding system and the planning system, and the exploitation of the technology market. The reform should overcome the defect of relying purely on administrative methods in science and technology management.

In the past, the government allocated research funds unconditionally to research institutes according to the number of staff, and was indifferent to the economic aspects of the research community. Under the reformed system, for a given period of time funds provided by central and local departments for S&T will increase gradually, at a rate higher than the growth in state revenue, in order to encourage the development of S&T activities in general.

For basic research and some applied research, funds will be provided through research foundations. But for research institutions engaged in technology development activities, it is planned that government contributions for current expenditure will be gradually reduced and abolished. Banks will be actively encouraged to provide loans for scientific and technological work, and to supervise and control the use of such operating funds.

Research institutions engaged in such important public matters as medicine, public health, labour protection, family planning, the prevention and control of natural calamities, environmental sciences, and other social sciences, as well as institutions providing certain scientific and technological services, will continue to receive state funds in accordance with existing block funding practice.

The commercialization of technological achievements and the exploitation of the technology market will be arranged to suit the development of the socialist commodity economy. It is realized that technology plays an increasingly important role in the creation of the value of commodities, and more and more technologies have become intellectual commodities in their own right. The intellectual industry has now emerged as a new trade. The technology market will therefore constitute an essential component of China's socialist commodity market.

The exploitation of technology covers seven main points and includes actively developing diverse forms of trade in technological achievements, technological job contracting, technical consultancy, and other services. Furthermore, the establishment of different kinds of business institutions dealing with technological commodities will be appropriately supported. Various measures will be taken to encourage enterprises to use new technologies and to improve their economic ability to satisfy buyers' demands. Statutes and regulations will be formulated to protect the legitimate rights and interests of buyers, sellers, and intermediaries. The ownership of intellectual property will be protected by the state through patent laws and other relevant statutes. The market prices of technological achievements will be determined through negotiations between sellers and buyers with no restrictions imposed by the state. All income from transfers of technology achievements will, for the present, be exempted from taxation. Technology development units and enterprises may reward personnel directly engaged in such development with a portion of the income from technological transfers. And units responsible for technological achievements may set up joint ventures with enterprises by contributing shares in the form of technologies.

Research projects having national priority will remain under the control of state planning, while other activities conducted by scientific and technological institutes will be managed by means of economic levers and market regulation, in order to enable these institutes to develop through internal impetus and to imbue them with vitality.

The reform of the organizational structure will change the situation whereby a disproportionately large number of research institutes are detached from enterprises; where coordination is lacking between research, design, education, and production; where the defence and civilian sectors are separated from each other; and where barriers are erected between various departments and regions.

The organizational structural reforms will be focused on strengthening the enterprises' ability to absorb and develop technology, and on strengthening the intermediate links in a complete life-cycle production system, as shown in figure 3. Emphasis is placed on encouraging partnership between research, educational, and design institutions on the one hand and production units on the other. Some of the detailed suggestions are as follows. The institutes of higher learning and research institutes under the Chinese Academy of Sciences, the central ministries, and local authorities will be encouraged to set up various forms of partnership with enterprises and design units on a voluntary and mutually beneficial basis. Some of the partnerships may gradually become economic entities. Some research institutes may develop on their own into enterprises of a research-production type or become joint technology development departments for small and medium-sized enterprises. Large, key enterprises would gradually improve their own technology development departments or research institutes. Defence research institutes would create a military–civilian partnership. While ensuring the fulfilment of national defence assignments, they would serve economic construction, accelerate the transfer of technology from the military to the civilian sector, and engage energetically in research and development programmes for civilian products. Collectives and individuals would set up research or technical services on their own. Local governments would exercise control over them and give them guidance and assistance. Institutes in this category would be profit-oriented.

The objective of the reform of the personnel system is to create a situation favourable to the emergence of a large number of talented people who can put their specialized knowledge to good use.

"The proper person in the proper position" is the optimum way to utilize human resources. The older generation of China's S&T specialists will be encouraged to continue to play their role in training qualified personnel and directing research, in writing books, and in acting as consultants to promote various public activities. A great number of accomplished and vigorous young and middle-aged people will be assigned to key academic and technological posts. Scientists and engineers in their forties and fifties will be able to contribute their full share as a bridge between the older and younger generations. Young talents will be nurtured.

In solving the serious "ageing problem" in the leading bodies of many research institutes, measures will be taken to train different types of scientific and technological managers, a new breed which possesses both modern scientific and technological knowledge and management skills.

Mobility of personnel will be encouraged. Competent people will no longer be made to sit idle and waste their talents. Appropriate S&T policies and preferential measures will be adopted to encourage S&T personnel to work in small and medium-sized cities, in the countryside, and in regions with communities of minority nationalities. Research and design institutes and universities will gradually experiment with recruiting personnel by invitation so as to break the so-called "iron bowl."

Active efforts will be made to improve the working and living conditions of scientific and technological personnel. The principle of "from each according to his ability, to each according to his work" will be earnestly adhered to in order to oppose egalitarianism. Rational remuneration for scientists and engineers will be gradually introduced. A system of honours and material rewards will be instituted.

The management system in agricultural science and technology will be reformed so as to serve the restructuring of the rural economy and to facilitate its conversion to specialization, commercialization, and modernization. In vigorously promoting technology development, efforts in applied research will be redoubled and basic research will be guaranteed steady and continuous growth. Opening to the outside world and establishing contact with other countries is a basic, long-term policy for China's scientific and technological development. Within this field, the utmost will be done to integrate foreign trade with technology and industrial production, and greater importance will be attached to importing patented technology, technical know-how, and software. More channels will be opened to expand various forms of international cooperation in development, design, and manufacture.

Certain domestic research and development work will be closely related to imports in order to absorb advanced imported technology. A policy of active support will also be adopted for technology development projects with promising international prospects so as to promote the ability of Chinese commodities to enter the world market. Active efforts will also be made to expand international academic exchanges.

The basic function of S&T in the coming decade is not to try to catch up with the advanced countries. S&T should first play its role in industrialization and serve the development of traditional industries and of technological rehabilitation. The industrialization process to be carried out will rest on the new S&T-based technological revolution.

In R&D institutes reforms will be carried out to strengthen the horizontal linkages between the universities, research institutes, and enterprises, between domestic regions, and between China and the external world.

R&D will have three functions. These are, first, to carry out strategically important basic research and applied research for selected areas of long-term importance, and, second, to develop and carry out a "technological complex" strategy, according to which priority will be given to the development of traditional technology (special importance being attached to energy, transportation, and the supply of raw materials, which are the bottlenecks in present economic development). Other priorities are the development of high technologies (with electronics and information technology as the guiding technology) and the reform of traditional industries and technologies, which will be combined with high technologies.

Thirdly, a "Sparks Programme" for the rural areas, where more than 75 per cent of the population live, will be carried out. Whether S&T can be effectively utilized in the industrialization process of rural areas depends on the sociocultural pattern of the regions. Guidance from the government plays a very important role in promoting this process. The "Sparks Programme" is sponsored by the State Commission of Science and Technology, which entrusts scientists and engineers with the job of designing simple and low-cost equipment with a high economic efficiency for rural use.

The S&T system must adapt to the structural changes of a dynamic society, and therefore human resource inputs must adapt to the dynamic changes of the S&T and economic system. China's present stock of educated manpower has an unusual composition. The proportion of the population with primary education is high, but the number of people with advanced educational qualifications is small (table 12).

This unusual structure has dual implications. It is favourable to grass-roots participation for self-reliance in S&T and economic growth. As the World Bank points out: "Widespread basic education in some East Asian economies has helped achieve both usually high output per unit of physical capital and, unusually, equal sharing of the benefits of rapid development." Yet, on the other hand, this type of educated manpower structure is unfavourable to the development of self-reliance in S&T, because a minimum number of highly educated people is required to promote the development of S&T. The present S&T specialized personnel structure in the national economy (table 13) is insufficient and could hamper growth even in the traditional sector.

In order to compare the relative intensity of specialized personnel employed in different branches of the national economy, a notion of "equivalent density" of specialized personnel has been adopted. This is defined as the percentage of the number of employees. A college graduate is taken as unity, while one who has completed graduate education, short-cycle higher education, or specialized secondary education is assigned an equivalent value of 2, 0.6, or 0.2 respectively.

Educational reforms entitled "Decision on the Organization and Management of the Educational System" were announced in 1985. The basic purpose of the reform was to produce more skilled, high-quality manpower. There would be nine-year compulsory schooling. The structure of secondary education would be readjusted and vigorous development of vocational technical education undertaken. The system of college admission, recruitment planning, and placement of graduates would be reformed and the autonomy of higher education institutions would be extended. The supervision and guidance of education would be strengthened, so as to ensure successful implementation.

Besides these five main steps, a multi-level and efficient system of higher education would be built up in order to remedy the shortage of highly educated manpower.

Social capacity in adapting new and existing technology is important in the management of S&T development. China already had a socio-

Table 12. Education attainment of population, by age and sex (percentages)

Country and age-group	Percentage of persons who have completed at least							
	Primary school		Lower secondary school		Upper secondary school		Post-secondary school	
	Male	Female	Male	Female	Male	Female	Male	Female
China								
15+ (total)	79.1	51.1	42.9	26.0	13.3	8.3	1.0	0.3
15–24	95.1	82.2	71.0	53.6	23.5	17.8	0.1	0.1
25–34	88.8	61.9	48.0	26.4	13.4	7.6	0.8	0.4
35+	63.2	24.6	21.5	7.5	6.4	2.4	1.6	0.5
India								
15+ (total)	37.2	14.7	21.3	7.1	10.3	3.0	1.6	0.4
15–24	53.6	27.4	35.5	15.1	15.6	6.4	1.2	0.7
25–34	39.4	14.4	23.4	6.5	13.1	3.1	2.7	0.7
35+	26.4	7.1	12.0	2.4	5.9	0.9	1.3	0.2
Republic of Korea								
15+ (total)	82.3	66.3	44.5	22.4	23.7	8.9	6.2	1.6
15–24	97.5	95.9	55.9	39.3	23.2	13.1	1.6	1.5
25–34	94.9	85.7	57.9	29.6	37.2	14.2	11.7	3.4
35+	63.7	37.7	28.2	8.2	16.3	3.6	6.6	0.7

Source: China: World Bank Country Economic Report, 1982.

Table 13. Percentage of specialized personnel in different sectors

Sector	%	Sector	%
Space	15.7	Weaponry	2.9
Shipbuilding	11.4	Building materials	1.99
Nuclear	10.5	Coal mining	1.87
Civil aviation	8.3	Urban and rural construction	
Electronics	7.3	and environmental protec-	
Petrochemicals	7.4	tion	1.83
Water conservancy and power		Textiles	1.58
generation	6.2	Highway and water transporta-	
Railways	4.6	tion	1.48
Non-ferrous metallurgy	4.4	Silk industry	1.36
Petroleum	4.3	Light industry	0.96
Ferrous metallurgy	4.2	Retail and wholesale trades	0.73
Chemical	3.8	Finance, banking, and	
Post and telecommunications	3.8	insurance	3.4
Machine-building	3.6	Jurisprudence and public	
Automotive industry	3.5	security	4.9

a. Survey carried out by the Ministry of Education from the statistical returns of 72 central departments.

infrastructure in planning; the problem was how to improve the planning system, i.e. to reduce the scope of mandatory planning and to improve indicative planning.

The long-term national development goal was to catch up with the developed countries by 2050 while maintaining a socialist system in which the benefits of prosperity are widely shared, avoiding polarization. Another goal for the year 2000 was to improve living standards and eliminate poverty, quadrupling the gross value of industrial and agricultural output (GVIAO) between 1980 and 2000 and increasing per capita national income from about $300 to $800. The leaders of the government have emphasized repeatedly the role of S&T in achieving these goals.

The strategy for development towards S&T self-reliance was, as explained earlier, to create a self-perpetuating mechanism for the S&T system through economic reform and the supportive sociocultural infrastructure. Both self-reliance and the transfer of technology from abroad will be emphasized, taking into account latecomers' ability to borrow. In the modern interdependent world, it is feasible for latecomers to adopt a "follower's" strategy but at the same time to increase their capacity to innovate and to adapt new technologies.

Because of different levels of development in different regions and different sectors, different strategies and S&T development policies will be adopted.

The resource allocation for science and technology R&D in China is roughly 1 per cent of its GDP. This is low in comparison with the R&D expenditure of many of the developed countries, such as the USA, United Kingdom, Germany, and Switzerland, where R&D expenditure generally exceeds 2 per cent of their GDP. Around 70 per cent of this outlay is allocated to development, while the remaining 30 per cent is divided between basic and applied research. The resource allocation for S&T and between sectors is entirely determined by the government according to its choice of priority sectors.

Nearly all R&D systems in the past were publicly owned. R&D management, in the words of one Western scientist, was such that

. . . despite a very elaborate structure of R&D functioning relatively well in China in the context of developing countries, the Chinese still do not have what may be called a systematic effort to build or utilize R&D management expertise on modern lines . . . that can be explained as a necessary consequence of not having so far developed general management education and training on a system basis. . . .

This point has now drawn the attention of the Chinese government, and reforms are being undertaken. But difficulties could arise because many components are involved in this reform, and, in the transitional period, interaction between the old system components and those of the new system has to be taken into consideration.

3

The Republic of Korea

Preamble

Self-reliance in science and technology (S&T) is defined here as the potential capacity to innovate and adapt either existing or new technologies. This definition assumes that a country has technology demands which vary according to its economic and social condition. The technology that is needed can be either obtained domestically or imported. In the latter case, the technology is bought because it is cheaper to import it than to generate it in the local environment. This also implies that a given society has the potential locally to obtain previously imported technology, at a production cost not very much higher than the import price, if importing turns out to be impossible.

A society pursuing self-reliance in S&T would also normally trade technologies with other societies, and may in fact have a trade deficit. The society maintains the deficit and begins to provide itself with the technology it needs.

The S&T in focus here includes not only technology but also scientific knowledge and know-how. In other words, self-reliance aims at achieving self-sufficiency across the whole spectrum, ranging from basic scientific knowledge, which is easily available from professional journals, to on-the-spot application methods that are imported if needed for use in a short span of time.

This definition of self-reliance in S&T assumes that developed economies are close to being self-reliant, while developing economies are not. But among developing economies, the degree of self-insufficiency may vary depending on the condition of the particular economy. A rapidly growing economy will have a lower degree of self-reliance than a stagnating economy, not because its supply capacity is limited, but because its demands expand faster than the available supply.

The acquisition of technology can be divided into four different stages as follows: (1) operation, maintenance, and repair; (2) imitation and modification of foreign technology; (3) design; and (4) mass manufacturing based on independent design.

The discussions below on the Korean experiences assume the above background of S&T self-reliance.

History

Korea's written history, stretching back over 4,000 years, has been influenced by Buddhism and Confucianism. Its "closed door" policy towards the Western world was abandoned in 1867 under pressure from Japan and other countries. Korea was now obliged to allow access to Western trade and technology. Further, starting in 1910, Korea was occupied by Japan for 35 years.

Japanese rule significantly influenced Korea. She suffered the consequential misfortunes of economic dependence and cultural repression, although the customs and structure of society were not deeply affected. Japanese influence was most evident in the administrative and legal fields. During this period, a substantial inflow of Japanese capital and administrators changed Korea's economic structure, particularly through the introduction of new agricultural practices and the development of a fairly significant industrial base. However, the potential impact of the Japanese was limited by the restrictions placed on Korean participation in economic and political planning and management.

Korea regained her independence in 1945, but was partitioned into two states along the 38th Parallel. This had far-reaching consequences. The division destroyed an interdependent economy, giving the South less than half the land, about three-fifths of the population, most of the agriculture, and very little of the industrial capacity of undivided Korea.

With the establishment of the Republic of Korea in 1948 in the South, some economic progress was made in the first two years of the new republic. The most significant development was the land reform programme, which redistributed about three-fourths of the cultivatable land and benefited over half the rural households.

Soon afterwards, the Korean War (1950–1953) devastated almost the entire country: large sections of the economy were destroyed and a quarter of the population became refugees. War deaths alone were estimated at over 3 million for both sides.

With financial assistance, largely from the US and the UN, the process of reconstruction of the ruined economy was undertaken after the war, and the foundations of the industrial sector were laid. During the next few decades, the economy notched a respectable rate of growth, mainly by the expansion of industry.

Development policies and strategies from the 1960s to the 1980s

Under Japanese domination, Korea was mainly an agricultural economy, exporting rice and importing manufactured goods. Since its liberation from Japanese occupation, the Korean economy developed in three phases: periods of economic instability and destruction from 1953 to 1954; reconstruction and expansion of the economy from 1954 to 1961; and accelerated economic growth after 1962.

During the reconstruction period, import substitution of consumer durables was attempted by guaranteeing a secure, though limited, market. Production was also expanded continuously throughout this period.

The foreign exchange required for industrialization during this period was limited. This was obtained through the export of agricultural goods and other raw materials, and partly through limited transfers from abroad. Because of this foreign exchange shortage, the Republic of Korea shifted, from the 1960s on, to an export-oriented growth strategy, exporting labour-intensive manufactured products.

Significant development and increases in real per capita GNP were seen only after the first Five-Year Economic Development Plan in 1962. From 1962 to 1984, real GNP grew at an annual rate of 8.2 per cent – the result of the adoption of the export-oriented industrialization strategy – and was much higher than the figure of 4.3 for the period 1954–1961. As the Korean economy became productive, foreign capital entered the country voluntarily, in the form of both loans and direct investment. This not only contributed to high growth rates, but also resulted in a high external debt.

Soon, the export of labour-intensive products reached its limits, as labour costs rose and the labour surplus disappeared. The economy now had to shift to other types of manufacturing with different comparative advantages. One was the import substitution of capital goods, which until then were imported.

This shift was not easy, as the technologies in capital goods manufacturing were much more sophisticated, and, in contrast to labour-intensive manufacturing, it was hard to attain scale production. A compromise was the choice of a limited range of capital goods that were relatively labour-intensive, as in the machinery industry, construction equipment, and shipbuilding. However, even here, the labour required was of a more skilled character. In pursuing this strategy of skilled-labour intensive manufacturing, the Republic of Korea faced shortcomings in both technologies and skilled labour.

To overcome these difficulties, the country now began to import and then adapt foreign technology whilst simultaneously enhancing its indigenous technological capability. In this difficult task it was broadly successful, although there were partial failures, as in the fields of precision machine tools and heavy electrical equipment.

The plans

One of the main development strategies and aims of the first Five-Year Plan (1962–1966) had been the formation of the social overhead capital for industrial development, as in electrical power generation and transportation, and the production of important industrial goods that support industrial development. Another main aim was the building of light consumer goods industries, principally the synthetic textile industry for both import substitution and exports.

The second Plan (1967–1971) continued and accelerated this thrust. Strategically important high-quality labour-intensive industries, such as electronic consumer goods and synthetic textiles, were selected for exports. During this Plan period, the foundation was also laid for heavy and chemical industries, with an emphasis on machinery, shipbuilding, electronics, iron and steel, non-ferrous metals, and petro-chemicals. The government also initiated the systemization of industries, the building of industrial complexes, and the cultivation of technical manpower.

The third Plan (1972–1976) saw a significant change in the Republic of Korea's industrial structure. Heavy and chemical industries grew rapidly, while the modernization and balancing of the industrial structure was promoted, indicating a new growth stage. The export of manufactured goods from these industries increased and new industrial expansion also occurred through the establishment of industries in various regions.

During the fourth Plan (1977–1981), a further diversification of the industrial sector was carried out, and technology-intensive industries

Table 1. Performance of Five-Year Development Plans

Growth rate (%)	First Plan	Second Plan	Third Plan	Fourth Plan	Fifth Plan[a]
GNP	7.8	9.6	9.7	5.8	7.0
Industry					
Agriculture, forestry and fisheries	5.6	1.5	6.1	−0.6	4.3
Manufacturing	15.0	21.8	19.0	10.5	11.5
Import/export					
Commodity export	38.6	33.8	32.7	11.1	14.4
Commodity import	18.7	25.8	12.6	10.5	10.6
Fixed capital formation	24.7	17.9	13.2	10.1	8.3

Source: Economic Planning Board (EPB), *Major Statistics of Korean Economy*.

a. Figures are from the revised plan.

such as machinery, electronics, and specialty chemicals were developed. The fourth Plan also expanded the export of production materials and capital goods. A long-term development strategy was formulated to promote technology and knowledge-intensive industries and to achieve the high-level industrialization characteristic of an industrially advanced nation.

The main objective of the fifth Plan (1982–1986) was to solve the problems caused by the rapid economic growth that had occurred since the 1960s, as well as to continue economic growth. Table 1 shows the performance of the different Five-Year Development Plans.

Impact on the agricultural and industrial sectors

In its development, the Korean economy showed a disparity between agricultural and industrial growth. This had several causes, among them the changes in the structure of foreign demand, the changes in domestic demand as per capita income rose, the policies that favoured industrialization for export, the uneven expansion of factor inputs, the unequal increases in productivity between the sectors, the different rates of adoption of new technologies, and the fact that the industrial sector received priority in the government's economic plans.

There was a wide disparity between agricultural and industrial growth of total output as well as output per worker. Table 2 shows the employment and output per worker in both sectors during the 1962–

Table 2. Sectoral employment and output per worker

Year	Agriculture and forestry			Industrial sector		
	Employment		Output per worker (thousands of won)	Employment		Output per worker (thousands of won)
	No. (thousands)	%		No. (thousands)	%	
1963	4,822	60.7	42	631	7.9	116
1964	4,906	59.8	63	671	8.2	166
1965	4,603	56.1	64	772	9.4	188
1966	4,695	55.7	74	833	9.9	232
1967	4,598	52.7	82	1,021	11.7	234
1968	4,582	50.0	93	1,170	12.8	280
1969	4,687	49.8	120	1,232	13.1	350
1970	4,826	49.5	141	1,284	13.2	436
1971	4,785	47.3	175	1,336	13.3	530
1972	5,110	48.4	200	1,445	13.7	635
1973	5,260	47.2	229	1,774	15.9	753
1974	5,304	45.8	327	2,012	17.4	948
1975	5,123	42.3	451	2,205	18.6	1,190
1976	5,323	42.4	570	2,678	21.3	1,412
1977	5,161	39.9	690	2,798	21.6	1,732
1978	4,920	36.5	903	3,016	22.4	2,213
1979	4,642	34.0	1,121	3,126	22.9	2,751
1980	4,433	32.3	1,095	2,972	21.7	3,602
1981	4,560	32.5	1,468	2,872	20.4	4,555
1982	4,324	30.0	1,611	3,047	21.1	4,755
1983	4,043	27.9	1,828	3,275	22.6	4,984
1984	3,726	25.8	2,201	3,351	23.2	5,688

Sources: GNP: Bank of Korea (BOK), *Economic Statistics Yearbook*, 1985; employment: EPB, *Annual Reports on the Economically Active Population*, 1985.

1984 period. The growth of output per worker has been substantial, for two reasons. One is intersectoral, a result of the changes in the distribution of employment as workers shifted to sectors where output per worker was high. The other is intrasectoral, a result of increasing output per worker in any given sector. The output per worker in the agricultural sector was high; yet the disparity, as a ratio, has remained roughly unchanged. During this period the distribution of employment shifted substantially from the agricultural to the industrial sector.

The heavy concentration of industry in large cities and industrial complexes probably helped the export effort, but also resulted in a rural–urban income disparity. Since the early 1970s, there has been an attack on this unbalanced regional growth. One important objective of new policies has been to create more industrial jobs in the rural areas and thus expand off-farm employment opportunities.

Science and technology in Korea before the 1960s

The introduction of Western culture and technology to Korea began around 1880. Until 1910, the introduction of modern technology was dominated by Japan and the Western powers. Some technologies in small arms, explosives, agriculture, paper, mining, printing, leather goods, communications, and railroads were introduced from Japan, the US, Germany, and Russia. Electricity was introduced from the US in 1898. The first scientific institution was the Industrial Research Institute, founded in 1883.

Under their rule, the Japanese introduced, between 1910 and 1945, Western technology on a large scale. During this period hydroelectric power, fertilizers, cement, textiles, and steel industries developed, along with a general consumer goods industry. But the development of industry was determined by the Japanese strategy for controlling the Far East and preparing for the Second World War, with a main emphasis on mining and transportation.

The production of crude and semi-processed agricultural and mineral products was intended for export to Japan and its other colonies. Manufacturing during the colonial period also had a heavy Japanese imprint with regard to the capital equipment, entrepreneurs, engineers, and technicians involved, and even the labour, particularly skilled labour.

Steady progress was made during this time in the education system. The first modern university was established under Japanese rule in 1924, while the first college of engineering was set up in 1938. However, a strict ratio between Korean and Japanese students severely restricted Korean participation. By 1945, only 800 Koreans had graduated, of which 300 were from the Medical School and about 40 from the School of Science and Engineering. After independence, education, including technical and higher education, was sharply accelerated. During this colonial period, Koreans acquired mostly on-the-job knowledge of the operation of modern industries.

Industrial development during the Japanese period, it should be noted, was concentrated in the North. So, when Korea was divided into North and South after the Second World War, there was almost no heavy industry in South Korea.

After the end of Japanese colonialism, the relationship with the United States served to augment Korean resources, both directly and indirectly, especially in the formation of human capital. American aid directly contributed to the rapid expansion of education, which by 1960 led to universal primary education and nearly universal adult literacy, while contributing to increasingly higher enrolment rates at all grade levels above the primary. Aid also financed the overseas education and training of thousands of Koreans.

An indirect contribution, because of universal military service, was made by American military advisers. The Korean military learnt modern concepts and techniques of management and organization. For the labour force, military service was an important source of skill formation, similar to training in modern industry. An important channel of industrial technology was technical advisers.

The role of science and technology in recent development

Beginning with the first Five-Year Plan, the Korean government enthusiastically promoted the development of S&T to support socio-economic growth. During the 1960s, Korea was completely dependent on advanced countries for production facilities and technology. The choice of the appropriate technology to adopt was a very important task for economic development. The choice of strategic industries for economic development was affected by the possibility of success in technological adaptation.

A primary emphasis was put on the import of advanced technologies for solving problems that arose in import-substitution industries, such as the energy, fertilizer, and cement industries, and in the export-oriented industries that were now being promoted. In 1960, a law was enacted promoting the inducement of foreign capital and this hastened the introduction of foreign technology and capital goods. But the accelerated importation decreased foreign exchange holdings and resulted in many unfavourable contracts.

Consequently, the government turned to a passive but stable policy on the import of foreign capital and technology. As a result, the number of technology imports decreased from seven in 1962–1963 to two in 1964, rising slightly to four in 1965. The major sources for technology imports were the United States and the Federal Republic of Germany.

In 1965, a Korea–Japan agreement led to resumed diplomatic relationships, and Korea obtained the right to demand compensation from Japan. As a result there was a significant ten-year inflow of Japanese capital, starting in 1966, which made Japan the major source of technology. The number of technology imports now increased sharply.

The primary emphasis of S&T policy during the 1970s was the adaptation and improvement of imported technologies and the establishment of private research and development (R&D) systems. During this period, joint investment of domestic and foreign capital was promoted, with a consideration for its spread effect on domestic technology. Foreign interests had made investments since 1962 and these had been favoured by the Korean government. But direct foreign investment turned out to be ineffective in the domestic dissemination of foreign technology. Therefore, the government enacted a law for foreign capital inducement which added some limitations on the terms and conditions for direct foreign investment while encouraging the import of technology.

The development of heavy and chemical industries in the 1970s rapidly increased the demand for advanced foreign technology. The Republic of Korea's foreign exchange holdings had increased remarkably from the mid-1970s, and the government realized that, for further economic growth, it was necessary to transfer its leadership in economic and technological development to industries. These factors, along with an international trend in liberalization of trade and foreign investment, forced the Korean government, from 1978, gradually to liberalize technology and capital imports.

In the 1980s, the country took another leap forward towards the goal of being an advanced industrialized country. In the achievement of this goal, S&T played an active part, leading, rather than supporting, economic growth. Considerable support was provided for graduate education, basic as well as applied sciences, and university research in basic science.

143

Science and technology and the exogenous environment

Because the import of foreign technologies requires foreign exchange, it has been controlled by the law for foreign capital inducement. Before the revision of 1978, every contract for technology imports was reviewed by the Minister of the Economic Planning Board. The operational decrees of the foreign capital inducement law, revised in 1978, classified technology imports under three categories, enabling some imports to be exempted from the review.

To be classified in the review-free category, the contract would have to be short-term and worth only a small amount and the imported technology would have to be deemed necessary for any one of the following industries: shipbuilding, machines, electronics, electrics, metals, chemicals, or textiles. Technology imports in other categories still had to be reviewed. Long-term contracts or ones involving large amounts and/or the import of technologies necessary for the nuclear, computer, or defence industries belonged to this latter category. Technology imports not included in either of these two categories were exempted from review only when no objections were raised within 20 days by the ministry from which the Minister of the Economic Planning Board asked an opinion.

More revisions were later made in the direction of liberalizing technology imports, and finally, in June 1984, individual review was completely abolished. As a result, technology imports nearly doubled, the amount spent annually increasing from $120 million in 1983 to $213 million in 1984.

Education and training

In the 1960s, the education and training of technology manpower had been led by the government. The light industries of this period required only a supply of enough technicians and some college-graduate engineers to be able to operate and maintain industrial facilities. Manpower policy during the 1960s, therefore, concentrated on training a sufficient number of technicians.

The industrial demand for technicians rapidly increased after the first Five-Year Plan, but school education could satisfy only 30 per cent of this demand. The government, therefore, enacted in 1967 a law on vocational education in order to encourage or force industries either to train the necessary technicians directly or to finance their training at vocational schools. In the same year, the government also enacted a law for job stabilization, in order to reduce frictional unemployment and utilize technical manpower efficiently.

In the 1970s, high-level technologies that were imported required a good base of R&D personnel for their assimilation. Hence, graduate education in science and engineering was actively promoted during the 1970s.

The establishment of the Korea Advanced Institute of Science (KAIS) as a postgraduate school in applied science and engineering was a turning point. KAIS led the nationwide upgrading of graduate education and contributed to the establishment of a mass supply system of high-quality scientists and engineers. To supply sufficient numbers of college-graduate engineers to the heavy and chemical industries, the government began in 1973 to foster specified departments in some universities. Such departments and universities were appointed with consideration to the industrial needs of their regions.

From the mid-1970s, it was widely understood that, for further economic growth, the Republic of Korea would have to compete with developed countries in high-technology industries. The government realized the importance of basic science as well as applied science in this competition. As a result, the Korea Science and Engineering Foundation was established in 1977, in order to support researchers in basic science.

Fostering high-quality scientists and performing basic research became urgent tasks for the new higher technology requirements. Increasing the quantity and quality of graduate education was one effort in this direction. In addition, the government operated overseas study, training, and research programmes in S&T. These programmes consist of degree, training, post-doctoral, and research courses. Since 1982, the Korea Advanced Institute of Science and Technology has been offering Master's and Ph.D. courses for researchers in government-sponsored research institutes.

Four science high schools were founded in 1984 and the Korea Institute of Technology was established in 1985 to supply scientists comparable with first-class scientists in advanced countries. The foundation of these schools is also a long-term strategy in the competition with developed countries in high-technology industries, which is essential for future economic growth.

To encourage basic research, the Ministry of Education has since 1979 supported the establishment and financing of basic research institutes in universities. But universities spent less than 15 per cent of total R&D expenditures during the 1980–1983 period, although they had more than 40 per cent of total researchers. This may indicate that basic research has not yet been sufficiently supported. According to a recent survey, the proportion of basic research expenditure to total research and development expenditures since 1982 has been around 17 per cent. To foster basic research, this proportion would have to increase by up to at least 20 per cent and the increase in the basic research budget should be distributed to universities more proportionately.

Research and development

A turning-point in Korean S&T was the establishment in 1966 of the Korean Institute of Science and Technology (KIST). It was founded to provide technological assistance to the heavy, chemical, and other export industries. Since then, through the 1970s, about 10 specialized industrial research institutes were established, many being spin-offs from KIST. In 1971, the Korea Advanced Institute of Science (KAIS) was established as a postgraduate school in applied sciences and engineering to supply high-quality scientists.

The financial and technical support of the United States made the establishment of KIST possible. But it was in 1969 that KIST actually began to perform R&D. All the research institutes established since then have followed the pattern of KIST, and the Korean government has also made great efforts to repatriate Korean scientists from abroad and to utilize them in adapting and improving imported technologies. To repatriate these scientists, special incentives were provided and the autonomy of research activities, as well as their financial support by the government, was assured.

During the 1970s, the growth stage of Korean industrialization, an emphasis was placed on fostering industries with higher-level technologies. The government had selected six strategic industries: steel, machines, shipbuilding, electronics, petrochemicals, and non-ferrous metals. KIST had performed R&D in these fields until the mid-1970s but could not effectively meet their massive future technology demands. In 1974, therefore, the government enacted a law establishing specialized government-supported research institutes which would provide technology assistance to strategic industries. In the second half of the 1970s, a series of such specialized research institutes was established, each institute being financially supported through a relevant ministry.

Some of the specialized research institutes were spin-offs from KIST and some were reorganized from existing public research institutes to

146

enhance the flexibility and efficiency of R&D activities. Until the 1980s, these specialized institutes would spend most of their efforts in establishing research systems, rather than in actually doing R&D. Serious efforts were made to improve imported as well as existing technologies, and to supply qualified scientists and engineers, educating and training them to meet the industrial needs of field-related problem-solving.

During the 1970s the creation of a favourable environment for scientific activities was another major objective. To achieve this, the general awareness of the importance of S&T was increased and "scientific thought patterns" promoted. Educational activities along these lines were also emphasized by the Korean government and educational institutions.

In the 1980s, the Republic of Korea had to compete with developed countries in some high-technology industries. To succeed in such competition, R&D systems in Korea had to be reorganized into a more efficient and harmonious whole. In 1980, therefore, the government merged 16 research institutes to create nine new ones. The Ministry of Science and Technology (MOST) took responsibility for coordinating the R&D of the reorganized institutes. To make such coordination effective, MOST supported the financing of all of these institutes except the Korea Ginseng and Tobacco Research Institute, which is financially supported by the Office of Monopoly.

Since 1982, national projects have been carried out to compete internationally in the development of high technology. The two main categories of national projects are Government Projects and Government–Industry Joint Projects. During the period from 1982 to 1984, 214 Government Projects were conducted and 38 billion won was spent on them. Three hundred and forty-five Government–Industry Joint Projects were carried out during the same period, financed by industry to the tune of 17 billion won. These projects have contributed significantly to the sharp increase in the R&D expenditures of government-sponsored research institutes. The government budget for national projects has been increasing, and reached 30 billion won in 1985 and 50 billion won in 1986.

Table 3 shows the financial support made to R&D institutes by the government during the 1982–1985 period. For each year during this period the total spent amounted to 35–40 per cent of the government's budget for the field of S&T – a proportion that is much higher than before the reorganization of R&D systems in 1980. If we include the budget of the government-financed research projects carried out by

147

Table 3. Financial support of research institutes by the government (millions of won)

	1982	1983	1984	1985
Korea Advanced Institute of Science and Technology	19,351	19,827	18,151	45,092
Korea Ocean Research and Development Institute	1,713	2,326	2,787	3,639
Systems Engineering Research Institute	2,734	1,904	1,233	1,625
Korea Advanced Energy Research Institute	15,237	19,174	19,170	21,445
Korea Institute of Energy and Resources	13,080	13,874	12,524	16,873
Korea Standards Research Institute	3,297	3,545	3,308	4,514
Korea Institute of Machinery and Metals	6,642	7,711	7,709	8,278
Electronics and Telecom Research Institute	20,242	22,836	20,526	22,775
Korea Research Institute of Chemical Technology	1,870	2,401	2,553	3,024
Korea Ginseng and Tobacco Research Institute	6,004	6,384	5,821	5,650
Total	90,177	100,034	93,787	132,921

Source: MOST, *Statistical Year Book of Science and Technology*, 1984.

these research institutes, the proportion in each year of the 1982–1985 period amounted to 42–49 per cent.

Reassessment of the policy and strategy

To attain an equitable distribution of the benefits of growth, the government since the 1980s has made serious efforts to improve living conditions in rural areas and raise the income of rural households. The efforts to improve rural living conditions include setting up bus routes in small farming and fishing villages and constructing nurseries and schools.

Table 4 shows the ratio of paved to total local roads and of rural to urban incomes of households from 1981 to 1984. To raise farm incomes, efforts were made to use arable land fully, increase farm mechanization, and promote floriculture and stock-breeding. To increase non-farm rural incomes, small and medium-size factories were

Table 4. Ratio of paved to total local roads and the ratio of rural to urban household incomes (%)

	1980	1981	1982	1983	1984
Paved/total local roads	–	11.0	12.8	18.1	22.3
Rural/urban incomes of households	84.0	96.6	103.2	102.8	99.9

Source: EPB.

built in rural areas. As a result, the income of rural households now approaches that of urban households.

To enhance the competitive power of industries, the government reduced its leadership in economic activities and its protection of the domestic market. To promote competition in the market, a fair trade system was established and trade liberalization expanded. To strengthen the competitive power of industries, the government supported small and medium-size firms.

As a result of these efforts to promote competition, R&D investment in industries increased sharply. Table 5 shows the ratio of S&T investment to total GNP from 1980 to 1986. The government has also made efforts to improve the financial structure of the industries. Loans to big enterprises have been restrained. As a consequence, the capital –asset ratio has increased since 1980.

Since 1980, then, the Korean government has implemented some new policies and complemented ongoing policies for the further development of science and technology. Securing enough investment and adequate manpower was now the most important task for S&T development. The government made serious efforts to encourage private investment, first by offering tax incentives. Though tax incentives for R&D activities had been offered in the 1970s, they were not sufficient to induce enough private R&D investment. In the 1980s some new incentives were added and existing ones were strengthened.

The remarkable increase in S&T investment and R&D expenditures is mainly due to the sharp increase in the contributions of the private sector. The share of the private sector in total R&D expenditure increased from 48 per cent in 1980 to 70 per cent in 1984, although the actual performance by industry is somewhat lower than these figures. The difference arises from the fact that industry gives R&D to government-sponsored research institutes and universities.

The Republic of Korea's gradual mastery of technology resulted

Table 5. Investment in science and technology (billions of won)

	1980	1981	1982	1983	1984
1. GNP	37,205	45,775	48,088	58,280	63,277
2. Investment in S&T	317	404	526	682	886
3. 2/1	0.85	0.88	1.09	1.17	1.46

Source: MOST, *Statistical Year Book of Science and Technology*, 1984.

eventually in technology exports. Such exports in fact began in 1976 with Korean companies participating in construction works in the Middle East. But performance is still poor and technology exports (excluding technical labour and service exports) declined between 1981 and 1984.

Achievements in industrial development

To assess mastery of technology, Korean industrial performance can be taken as an indirect indicator.

Thus, in steel, the Republic of Korea developed from small-scale production with electric furnaces in 1960 to being one of the largest steel-producing countries in the world by the late 1970s, with a capacity of 13.54 million M/T of ingot in 1985. Beginning in 1973, the petrochemical industry achieved self-sufficiency – about 80 per cent of imports had been substituted – and had an export potential of downstream industries.

The electronics industry recorded phenomenal growth in a little over two decades, with the following landmarks: black-and-white television in the late 1960s, colour television in the early 1970s, and other consumer and industrial electronics in the mid- and late 1970s. In 1985, the export of electronics reached US$4.4 billion, representing 15 per cent of the total.

Up to the mid-1970s the shipbuilding industry in the Republic of Korea was small, but by the end of 1985 a maximum capacity of 1 million DWT was reached, placing the country in the world's top rank.

Automobile assembly began in the mid-1960s with completely knocked-down components. Real mass production began only in the mid-1970s, with 200,000 passenger vehicles manufactured in 1979 and 30,000 exported. Production capacity gradually increased to 685,000 per annum in 1985, with a local content ratio of 90 per cent. The automobile industry has now grown to be one of the important export industries, North America being a principal market.

The electronics industry as a case-study

As part of the research, two case-studies, of the electronics industry and the farm sector, were undertaken. As the electronics industry in the Republic of Korea is very significant, the main highlights of the case-study will be given here.

The Korean electronics industry has gone through five different phases: (1) random exploration; (2) technology introduction; (3) equity joint venture and Original Equipment Manufacturing (OEM); (4) copying and transforming; and (5) the development of its own technologies and products.

In the first stage of random exploration between 1958 and 1962, vacuum radio tube radios, electric fans, and telephone manufacture were begun. The production of these items was carried out with inadequate technology resources. For design, heavy reliance was placed on operation and maintenance manuals, circuit drawings, components lists, and schematic layouts that came with imported electric equipment. It was virtually impossible to obtain materials or information for manufacturing technologies. The manpower for manufacturing during this stage was drawn from those who ran minor repair shops, testers of telephones, and "dabblers" in electrical equipment.

During the second phase from 1963 to 1968, foreign technologies were brought in to fill the perceived inadequacy of the earlier efforts. The packaged deals that were entered into at the time brought in mass production. But such deals often meant that a foreign party was entirely responsible for selecting and supplying the manufacturing equipment and the technical assistance.

Laws to induce foreign capital were now set up, but encountered many difficulties in their implementation, including official inexperience and inefficiency. Yet import substitution was carried out and at the end of this stage about half the electronic and electrical manufactures were being exported. The major products manufactured included refrigerators, watt hour meters, telephone switching equipment, transistor radios, cables, and elevators. Other manufacturing attempts included communications, electronic home appliances, and consumer electronics.

The technologies during this second stage came from drawings and specifications supplied by foreign agencies, instructions given by foreign experts, and know-how brought back by Koreans trained abroad. An important manpower consideration at the time was the training of a large amount of subgrade technical manpower for the electronics industry.

The major technological concerns during this period were the manufacture of the moulds, jigs, and tools for mass production, prefabrication of mechanical parts and the production of simple passive circuit elements, and their final assembly and testing. However, foreign collaboration at this time precluded mastery of design, as well as the import of design equipment.

During the third stage of equity joint ventures and OEM (1969–1975), certain problems associated with mass production had to be faced. As the local market was limited, overseas markets were sought. But overseas marketing by Korean industry alone was found to be impossible. Because design capacities were not transferred, every time a new model was introduced fresh contracts had to be signed with the foreign partner. Large-scale manufacture also required advanced business skills beyond mere production and testing.

In spite of these difficulties, the number of technical collaborations increased for some time. Overseas markets were cultivated and management know-how was acquired. To attain these objectives, joint ventures, in which production management is partially carried out by the foreign partner, were increased in the fields that required more complicated manufacturing and management techniques, namely industrial and communications equipment. On the other hand, OEM was resorted to in the area of electronic home appliances for export.

With regard to joint ventures, the technologies acquired included manufacturing technologies that were slightly more sophisticated than those acquired in the earlier period and manufacturing techniques learnt through daily contact with foreign experts. No design technologies were transferred.

In the case of Original Equipment Manufacturing (OEM), quality control and testing techniques improved remarkably, partly through the process of learning from the foreign experts sent to do the final stages of testing. The comparative study of many different foreign companies proposing OEM also led to improvements in design capabilities. Unlike mere technology agreements, OEM involved the foreign party directly in the production fortunes of the company, so that technologies beyond the contractual ones were often transferred.

The new products manufactured during this stage included colour TVs and cassette radios. Progress was also made in other areas, improving models and increasing indigenization. The mass production of several circuit elements in joint ventures was also initiated. Several foreign firms began independent projects, assembling and manufacturing other electronic parts utilizing the cheap labour available.

As the positive effects of the electronics industry began to be perceived and its export potential realized, additional measures were made to foster it. These included the waiving of import duties on imported inputs for export electronics. The package of government supports helped strengthen the structure of the industry, rendering it partially immune to unfavourable external changes.

The fourth stage in the industry was that of copying and transforming (1976–1980), and this became necessary because of several changes. The oil shocks of the 1970s created a demand for electronically run control devices that would save energy. It also favoured less energy-consuming industries such as electronics. On the other hand, electronics-based automation also eroded the advantages of cheap labour on which some of the OEM prefabrication and assembly industries were based.

Furthermore, the shortening product cycles of new products were making the introduction of a particular technology for each model unattractive; and as the Republic of Korea's exports continued to increase, foreign collaborators were increasingly reluctant to transfer new technology. Many direct foreign investment and OEM ventures now virtually closed down overnight. And joint ventures aimed at the local market also came more or less to a standstill as local market limits were reached.

Because of these factors, Korean industry felt acutely the limits of the prevailing strategy of technology acquisition. It now began to venture out on its own by copying foreign products and transforming them to avoid legal and ethical complications. However, it was soon realized that, to copy foreign products successfully, custom-designed key components were required.

Major technological developments were reported during the period as industry responded to the changed circumstances. These included: increases in design capacity to produce models different from the ones being copied; the ability to design sequential control circuits using a microprocessor; the design of printing mechanisms for the Korean-alphabet office equipment; and the design of parts for electronic

PABX. In manufacturing, mechanical products capability, for example in audio deck mechanisms and Korean-alphabet mechanisms, was achieved.

These advances were aided by an increase in the stock of relevant technical skills. The latter included repatriated Korean scientists and engineers, graduates of KAIST, those trained in industry and private technology centres, and graduates of local engineering colleges.

The government also laid down new guidelines for the industry. Measures to help the industry included tax privileges and exemption from military service for electronics specialists, the latter permitting the release of adequate manpower for R&D efforts.

New products during this period included colour TVs, audio components equipment, and car radios in consumer electronics; typewriters and cash registers in electronic office equipment; and key telephone systems, electronic PABX, PCM carriers, and optical fibre cables in telecommunications equipment.

The last stage in the electronics industry involved the development of its own technologies and products (1981–). The continuing export of Korean electronic products had alarmed overseas markets and the licensing of foreign technology was becoming more difficult. The life cycle of products was getting shorter and shorter and a new model would appear in the market before significant progress could be achieved, thus blocking genuine local integration of the product. Creating new models by copying was also difficult because of the problem of custom-designed key components, which were impossible or very expensive to obtain from the respective foreign country.

When the local production costs of OEM or direct investment projects exceeded the level that foreign parties were willing to pay, they would suspend production. This immediately damaged the image of Korean products in the foreign market, and also tied up already invested facilities and technologies. Consequently, the crucial importance of self-reliance was realized.

This realization was reflected in the integrated circuits (ICs) that were developed during this period. An independent process of full-scale market cultivation was also launched to promote the overseas sales of VCRs. To reinforce existing know-how, the electronics industries established their own development centres to cover those technology areas for which no direct support was possible from other sources. They also established development centres in selected overseas countries, employing locals and available Korean engineers.

The Korean government also subsidized the incorporation of a number of research institutes apart from KIST, including the Korea Institute of Electronic Technology (KIET) and the Korea Electronic Telecommunications Research Institute (KETRI). In partnership with private industry, they developed computers and their peripheral equipment, the Very Large Scale Integration of electronic circuits (VLSI), and new materials.

In the case of the VCR, a successful though imperfect solution in basic design and mass production was achieved without external assistance. Considerable progress was also reported in mass production technologies, in quality stabilization using the most advanced automated equipment, and in the design of a process flow to make the best possible use of the relatively low wages in the Republic of Korea. The result of all these measures was that the Korean electronics industry had become a leader in consumer electronics and electric home appliances, and was now also establishing its own manufacturing bases overseas.

Self-reliance targets at each stage

The electronics industry has contributed much to the attainment of self-reliance in the economy and in science and technology, since it has achieved the fastest growth rate of all manufacturing industries.

This search for self-reliance occurred in four stages. In the first technology introduction phase, several foreign-oriented approaches were taken: the introduction of foreign loans for manufacturing facilities, the conclusion of licence agreements for production technologies, and, through the latter, the obtaining of licence drawings for product design.

At the second stage of equity joint venture and OEM, economies of scale were pursued. The OEM set-up was used to expand the market and the acquisition of managerial skills was achieved by equity joint ventures. During this stage, self-reliance was sought in relation to the foreign exchange burden. The industry had to service foreign exchange from the previous stage, while finding new sources of foreign finance to procure new parts and equipment. To achieve these ends, foreigners' own investments were encouraged, and were preferred to commercial loans. Thus, in order to be self-reliant in foreign exchange, the Korean electronics industry became reliant on foreigners for managerial skills in equipment production and in overseas marketing.

The next stage of copying and transforming was one of self-reliance with regard to the contractual assistance arrangements with foreign

155

companies. Unless this was reached, the industry would be unable to free itself from foreign restraints on decisions about the import of basic supplies and the independent cultivation of overseas markets.

This awareness prompted the industry to engage in product design and manufacture on its own without relying on manufacturing licences or on foreign capital. However, all the industry could do at this stage was to copy foreign products, transforming them as much as possible. This gave rise to two sets of difficulties. One was increasing regulation and restriction in intellectual property and the second was the importance of key parts designed by industrially advanced countries.

Overcoming these problems brought the Republic of Korea to the fifth stage, that of developing its own technologies in design and manufacturing. The industry would also continue seeking financial self-reliance by enhancing the value added on domestic products and promoting self-reliance through a build-up of its own overseas sales network.

Looking to the future, the industry's main necessity today is to free itself from reliance on foreign sources, principally Japan, for parts and raw materials. These and other similar problems have to be seen in the context of the need for a properly scheduled and systematic technology transfer. The need to identify and strengthen, for mutual advantage, the International Divisions of Industry is therefore pressing.

Problems and issues

In the 1980s S&T capacity in the Republic of Korea increased significantly. Yet the country is still far behind advanced nations. Its overall technology capability index (table 6) is only 5.7, while that of the US is normalized at 100. Thus, it is far behind the 86.9 of Japan or the 45.0 of the Federal Republic of Germany.

Yet S&T investment in Korea has increased faster than the GNP; the ratio of the former to the latter has steadily increased from 0.26 per cent in 1965 to 1.46 per cent in 1984. Investment has grown faster in the private than in the public sector. The share of the private sector in total R&D expenditure has increased from about 10 per cent in the mid-1960s to 79 per cent in 1984. Despite this remarkable increase in the past 20 years, S&T investment is still much less than that of advanced countries, and more investment is required.

The increase in manpower has been significant during this period. The number of researchers per 10,000 has increased from 0.7 in 1965 to 9.1 in 1984. Yet this figure is still much smaller than that of ad-

Table 6. International comparison of technology indices (1982) (billions of dollars)

	Number of patents and registrations of new design (1)	Technology trade (2)	Value added in manufacturing (3)	Export of technology intensive goods (4)	$\frac{(1)+(2)+(3)+(4)}{4}$
USA	57,889	7.5	642.3	109.2	
	(100.0)	(100.0)	(100.0)	(100.0)	(100.0)
Japan	105,905	2.3	321.1	91.4	
	(182.9)	(30.9)	(50.0)	(83.7)	(86.9)
Federal Republic of Germany	16,306	1.4	244.5	103.5	
	(28.2)	(18.7)	(38.1)	(94.8)	(45.0)
UK	29,590	1.8	124.9	42.3	
	(51.1)	(24.4)	(19.4)	(38.7)	(33.4)
France	23,944	1.3	154.3	43.3	
	(41.4)	(17.1)	(24.0)	(39.6)	(30.5)
Republic of Korea[a]	4,512	0.3	21.1	8.7	
	(7.8)	(3.7)	(3.3)	(7.9)	(5.7)

Source: Korea Development Bank.

a. Korean data are from 1983.

vanced countries. In addition, Korea still lacks high-quality, experienced engineers in the field.

The insufficiency of basic research is another serious problem. A recent survey indicates that the proportion of basic research expenditure to total R&D expenditure has been around 17 per cent since 1982. Serious support of universities by the government is required to increase this figure.

The main type of technology in the Republic of Korea still consists of simple processing and assembling technologies. One of the most serious problems in the field of S&T is the inferiority and insufficiency of the basic technology necessary for system design and the production of parts and materials. Improvement of these lagging basic technologies is critical for the upgrading of the industrial structure to a technology-intensive one. Because such upgrading is important for further economic development, greater efforts should be devoted to improving basic technology.

Future plan for self-reliance of science and technology

New technologies such as information technology and biotechnology are transforming the structure of industrial countries, and information technology is replacing mental labour by humans. Rapid progress and wide application of genetic engineering is expected to open a new era of prosperity for the world in the twenty-first century. Another breakthrough is the development of new materials such as fine ceramics.

These technologies are being developed at a very fast rate. The innovation rate for previous technologies did not generally exceed that of a business cycle, but the speed of major modern technological innovations, particularly semiconductors and computers, is often faster than this, causing an observable disturbance in market equilibrium. The implication is that technology can no longer be treated as a residual variable in understanding the economic behaviour of relevant industries. It is now to be understood as a major factor in economics.

There are also important differences in the development of such technologies. First, the life cycle of technological innovation, from basic research to product development, has been considerably shortened. Second, major innovations are now often the work of teams of many researchers. Third, the technology requires a multidisciplinary approach integrating many related "unit technologies."

The new technology is also changing the mass production system to one where many kinds of goods are produced in small quantities to meet the changing demands of consumers. Owing to the high risk and large investment involved in the new technologies, it is not likely that one country will become the leader in a wide spectrum of important technologies.

Through the application of high technology to existing industries, such as the automation of the textile and garment industries, developed countries will probably regain their comparative advantages in labour-intensive industries. Consequently the international division of production will no longer be based upon labour and capital alone, but on a new foundation centred around technology.

In this rapidly changing technological environment, developing countries such as the Republic of Korea are left with the very narrow option of developing a few selected technological areas to the level of advanced countries. For this reason, Korea has prepared the report "Long-term Perspectives for Science and Technology Development to the Year 2000."

The long-term goals and strategy of national development

In the next 15 years, the Republic of Korea's major goal is to make a smooth transition from a newly industrializing country to an advanced society. The advanced society visualized in the report has the following interrelated characteristics: freedom and stability; affluence and vitality; and justice and balance.

In order to meet the challenge of becoming an advanced society, its characteristics and traits, such as a strong motivation to learn, hard work, and determination to overcome difficulties, need to be continuously sustained and encouraged. The country must learn from the experiences of advanced countries and must exploit the advantages of a latecomer to development. We should carefully examine and analyse the experiences of advanced countries, in order to avoid the trials and errors that arose in the course of their development and to transplant successfully their policies and institutions into the Korean socio-economic climate.

Role of science and technology for future development

The role of science and technology in a future Korean society may be broadly stated as one of meeting felt needs by technological innovation and scientific advancement and of realizing long-term national goals for the next century. These goals fall under the following six headings:

To ensure national security and social stability

For a resource-poor country like the Republic of Korea, conserving energy and oil-substitutable energy is vital. Food technology is similarly important to maintain social stability. Science and technology are expected to play a vitally important role in ensuring national security and social stability.

To sustain the growth of the national economy and to improve its efficiency

In the past, technological progress made only a minor contribution to the growth of national income, and this should be changed. Furthermore, the technological gap with developed countries should be reduced in certain strategically selected areas. At the same time, technology intensification should be undertaken in small and medium-scale enterprises, which take approximately 97.5 per cent of the total, but contribute only 36 per cent to the Korean economy in terms of value-added production.

To prepare for a smooth transition to an information society

Societal change to an information society will demand the development of information-related industries centred around micro-electronics, communications, computers, etc. Moreover, reducing the labour component of production systems through automation technology will require re-education of displaced labour.

To improve the quality of life

Technology in areas of public health such as disease control, medicine and medical electronics needs to be developed. Another area is the protection of the environment for better dwelling conditions on the one hand, and for increased productivity of the land on the other.

Development of information technology directly related to daily living, it should be noted, will increase social benefits, and this in turn will help reduce urbanization. The preference for urban living will disappear with the development of an information system on a nationwide scale.

To create a new culture suitable for the new society

A conflict between traditional cultural values and progressive contemporary values has existed in Korean society during the recent process of industrialization. A national consensus should be created for the development of science and technology. Another far-reaching goal of science and technology is the creation of a new culture for the next century.

Long-term goal of S&T development

The long-term goal of science and technology should be in accordance with that of national development. The national development goal is stated as achieving equal ranking with the developed countries by becoming the world's fifteenth in terms of GNP and the tenth in terms of trade volume. To compensate for the country's paucity of natural resources, the necessary goal for S&T is to become no. 10 in the world in the area of industrial technology.

Because of the limitation in available resources, priority areas should be established through consideration of, among other things, national needs and comparative advantage. The role of S&T is to lead national development and to support socio-economic needs. The priority areas that have been identified are:

- Development of electronics, information, and communication technologies.
- Development of selected high technologies to lead the industrial structure adjustment.
- Development of key technologies to increase the international competitiveness of existing Korean industries.
- Development of technologies related to resources, energy, and food for social and economic stability.
- Development of technology in the area of health care, environmental protection, and social information systems to improve the quality of life and social benefits.
- Fostering of creative basic research to promote scientific advancement and to expand sources of technological innovation.

These priority areas were identified using the following basic criteria:

- Economic return and growth potential in view of limited development resources.
- Probability of success in view of development capability and experience.
- Indispensability in relation to national security and socio-economic stability.
- Industrial and technological linkage.
- Future contribution in relation to public welfare and new industrial possibilities.

The report "Long-term Perspectives for Science and Technology Development to the Year 2000" has laid down a set of policy guidelines in accordance with these basic directions, covering manpower, investment, national R&D systems, technological information systems, financial and tax support mechanisms, generation of a market for new technology products, technology intensification of small business, formation of R&D estates, build-up of a technology-oriented social culture, and internationalization of science and technology and international cooperation. Among the major recommendations were the following:

Manpower development will increase the proportion of scientists and engineers to the level of 30 persons (from the current 8 persons) per 10,000 population, amounting to about 150,000 scientists and engineers for R&D. Investment in R&D will be expanded from 1.7 per cent of the GNP in 1985 to over 3 per cent. The government and public sector will be responsible for 40 per cent of the total, and the private sector for the remaining 60 per cent, for which inducement policy instruments will be improved. Top priority in investment will be given to areas of high expected return, high linkage and externality, indispensability, and high probability of success and public interest.

For an efficient division of R&D activities, industrial firms would devote themselves to industrial technologies, whilst national and public institutes would be responsible for mission-oriented, applied research and for national projects of high risk and externality, and universities for basic research and manpower development, as well as for cooperation with public sector institutes and industries.

An efficient nationwide system of scientific and technological information will be established to collect, manage, and distribute information. Financial and tax support systems and public procurement will be used to induce the private sector.

In the next century, small and medium-scale industries will grow to form the backbone of the national economy, because the economic structure will change to one of "economies of variety" rather than "economies of scale." Therefore, support systems involving financial, tax, information, and public procurement measures will be specially devised for the technology intensification of small and medium-scale industries.

In the 1990s, in order to integrate education and research, particularly for high technology, R&D estates will be constructed in various regions to form a network. Development of technologies at regional level will contribute greatly to the balanced development of the Republic of Korea.

Summing-up and regional cooperation

In order to modernize the economy, the Republic of Korea adopted an unbalanced growth strategy for industrialization. Anticipated imbalances became apparent in many areas, for example between urban and regional development, between large-scale and small-scale businesses, and between export and domestic industries.

To redress the urban bias, increasing income in rural areas has become an important issue for the country. To this end, the mechanization of farming has been recently promoted in order to increase agricultural productivity. Opportunities for off-farm income generation have also been expanded through the creation of small-scale industrial estates in rural areas.

Large-scale enterprises have reached the level at which they can take care of themselves, and hence assistance in recent years has been directed towards the promotion of small and medium-scale enterprises. As a result, the latter are increasing both production and exports. This effort will be continued.

In order to balance export and domestic industries, a programme of industrial structural adjustment has been devised to effect a shift in the direction of technology-intensive industries. Here technology development emerges as most important. At their present level of development, Korean industries should possess enough indigenous capability to overcome the growing protectionism in technology and to increase their bargaining power for its importation.

Regional cooperation

At the present stage of development, the Korean economy can only maintain its sustained growth through ambitious internationalization.

163

Internationalization is not only desirable for domestic reasons, but is also in line with the growing interdependence of the world economy. As a leading NIC, the Republic of Korea is now expected to play an important role by both developed and developing countries.

The centre of the world economy is predicted to move to the Pacific region in the foreseeable future. In this region, Japan has excelled in its economic performance. For its own economic prosperity and for the sake of neighbouring countries, Japan is required to assume an appropriate role as economic leader.

The Republic of Korea will take an active part in internationalization, pursuing self-reliance in science and technology only in accordance with principles of interdependence. This interdependence will be realized in open competition and complementary cooperation. In this way, the country will both contribute to and benefit from the forthcoming Pacific Era.

At the same time, NIC and developing countries in the region should also be prepared to play a role. These countries have great potential if they can coordinate their efforts. With its relatively fresh experience of industrialization, the Republic of Korea may be able to provide some important lessons for countries with similar development goals. Cooperation with those countries will yield mutual benefits.

Japan is expected to be a valuable source of advanced technology for the future of the Republic of Korea, and technological cooperation will prove beneficial for both. Oddly enough, without such cooperation, Japan will also find it difficult to assume the future technological leadership of the world.

A standing regional organization will be useful not only for the efficient exchange of technical information, but also for technology implementation among countries in the region. Such an organization could review policies for technology transfer. Equally important is the establishment of a training centre in science policy and research management. Productive cooperation can be accomplished by the nurturing of mutual comprehension and a common awareness.

Bibliography

Choi, Hyung-sup. *Bases for Science and Technology Promotion in Developing Countries*. Tokyo: Asian Productivity Organization, 1983.

Economic Planning Board. *Current State of the Korean Economy and Economic Tasks for the Late 1980s*. Seoul: Economic Planning Board, 1988. (In Korean.)

Hasan, Parvez. *Korea: Problems and Issues in a Rapidly Growing Economy*. Baltimore, Md., and London: Johns Hopkins University Press, 1976.

Hong, Wongtack, and Anne O. Kruger, eds. *Trade and Development in Korea*. Seoul: KDI Press, 1975.

Kim, Kwung-suk, and Joon-kyung Park. *Structure of Economic Growth in Korea 1963–1982*. Seoul: KDI Press, 1985.

Kim, Kwang-suk, and Michael Roemer. *Growth and Structural Transformation (The Republic of Korea: 1945–1975)*. Cambridge, Mass., and London: Harvard University Press, 1979.

Korea Advanced Institute of Science and Technology. *Development towards the Year 2000*. Seoul: KAIST Press, 1986. (In Korean.)

Korea Development Institute. *Long-term Prospects for National Development towards the Year 2000*. Seoul: KDI Press, 1986. (In Korean.)

Kuznet, Paul W. *Economic Growth and Structure in the Republic of Korea*. New Haven, Conn., and London: Yale University Press, 1977.

Ministry of Science and Technology. *A Study on Science and Technology Development Systems and Its Future Direction*. Seoul: KAIST Press, 1986. (In Korean.)

Park, Chong-kee. *Macroeconomic and Industrial Development in Korea*. Seoul: KDI Press, 1980.

Park, Soon-jung, and Chung-soo Kim. "A Study on Technical Innovation and Our Countermeasures." *Korea Development Bank Monthly Economic Review*, April 1986, pp. 1–24. (In Korean.)

Wade, Larry L., and Byong-sik Kim. *Economic Development of South Korea: The Political Economy of Success*. New York and London: Praeger Publishers, 1978.

Woronoff, Jon. *Korea's Economy: Man-made Miracle*. Seoul: Si-sa-yong-o-sa Publishers, 1983.

4

Thailand

Thailand has taken considerable steps in national development since the country's first Social and Economic Development Plan in 1961. However, the country depended for this development upon imported production items and industrial technologies, which caused it to run up a high foreign trade deficit. In agriculture, moreover, the increase in the use of modern technology did not in general compensate for the high cost of production, with the result that the country's farm population was drawn in to an economic vicious circle.

The government has consequently been criticized for this distorted development, and for its poor strategy in the use of science and technology (S&T) for national development. A call for the appropriate use of S&T has been widely made across the country.

The present study develops a concept of self-reliance in S&T for national development, adapted from the successful examples of other countries. The major objectives of the study are: (1) to formulate a desirable strategy for national S&T development, and (2) to get an indication of the country's level of self-reliance in science and technology.

On the basis of these two objectives, the scope of the present study is to define self-reliance in S&T appropriate to the country's present social and economic capacity; to develop a conceptual framework for analysing the country's status in S&T; to carry out case-studies at both macro and micro level to give confidence in the framework developed; and to formulate an appropriate strategy for the development of science and technology.

The term self-reliance means different things to different people,

166

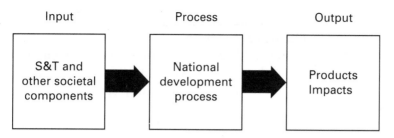

Fig. 1. Role of science and technology in national development

depending on the history of national development, and the social, economic, and cultural constraints of a country. For Thailand, self-reliance in S&T is defined as "the ability of the country to make autonomous rational decisions on science and technology in developing, selecting, implementing, managing, and operating technology and in replicating a useful technology in such a way that the country benefits."

There are five discernible stages in the development of technology, namely operational, adaptive, replicative, innovative, and creative. The study team holds that the first three stages are achievable whilst the last two seem beyond the country's capacity, given existing constraints.

Science is the knowledge of natural phenomena. Technology, on the other hand, applies a knowledge of science to help meet human needs and solve problems. Technology has two major components, software and hardware. The former concerns knowledge, while the latter constitutes the equipment and material used in production. S&T influences societal development and its role is represented in diagrammatic form in figure 1.

The above functional component inputs are interdependent: a change in one component creates a change in others. Science and technology is only a part of the process of societal development.

Traditional path of development

Thailand has been perceived as a "soft culture" society. Thais are noted for being very tolerant of the cultural and other differences of other peoples, and have absorbed many external cultural influences, particularly religion – principally from India – building technology, and, more recently, high technology.

Religion may be thought of as "software" technology, although, in the form of religious buildings and images, it could also be considered as constituting technological hardware.

In building technology, Thais have adapted their indigenous timber technology to create a distinctive Thai building style. They have developed this technology to the point that they can now use it to prefabricate houses.

The transfer of higher industrial technology, for example in transportation (railways, automobiles, etc.), started during the reign of Rama V when trade with European countries increased and modernization began. Other technological practices, such as the use of farm machinery in rice production, also began during the same period. Agricultural and other forms of production also changed from ones aimed at meeting only household and local demands to ones targeted at the export market. Modern irrigation was developed to support this new drive, and other necessary technologies, such as post-harvesting techniques, rice mills, river transportation, and food processing, were introduced.

After the Second World War, more automobiles, trucks, and aeroplanes were used in transportation. These accelerated the transfer of related technologies in, for example, road, highway, and railway-building technology. These, in turn, changed the educational curricula as the nation responded to these new technologies.

The newly introduced technologies influenced the modernization of the country beyond agriculture. Statistics show that the country's economy has experienced a satisfactory level of growth in production. However, doubts about the system have persisted, as some economic indicators, such as the trade balance and benefit–cost ratios, have shown negative trends. It appears that the more the country invests in production inputs and technology, the less the proportional output.

Development of the country in the national plans (1961–1986)

The country has been developed by three decades of national development plans, the first of which began in 1961. External political pressure, as well as the need to rebuild the country after the Second World War, and the demand for the elimination of inequality in income between the urban and rural sectors, influenced the development effort.

The first National Plan resulted in tremendous changes in the country's infrastructure, including increased transportation, roads, and railways, and a rise in the number of educated people. Income generation was also a focus. The figures given in table 1 suggest, at first sight, that

Table 1. Changes during national development plans (percentages)

Development sector	1961	1985	% change
Population			
Urban (%)	12.5 (1960)[a]	18.2 (1982)	+5.7[b]
Rural (%)	87.5 (1960)	81.8 (1982)	−5.7
Education			
People with basic education (%)	51.3 (1960)	82.4 (1980)	+31.1
People with higher education (%)	0.6 (1960)	2.3 (1980)	+1.7
Economy			
GNP at 1972 prices (millions of baht)		309,122 (1982)	–
Per capita GNP at 1972 prices (baht)		6,375 (1982)	–
GINI coefficient	0.5627 (1963)	0.6079 (1981)	4.52
Trade balance (market prices, in millions of baht)	−290 (1961)	−69,984 (1984)	
Infrastructure			
Roads (km) (1983)		33,148	
Railways (km) (1983)		3,735	
Airways in distance flown (km) (1983)		54,644,936	
Schools (no. per capita) (%)	0.1 (1961)	0.15 (1981)	+0.1
Land resources			
Agricultural area (%)	21.29 (1961)	45.83 (1984)	+24.54
Forest area (%)	53.33 (1959)	30.55 (1982)	−22.78

a. Figure in parentheses refers to year the data were obtained.
b. + means a quantitative increment only, not an improved quality.

the country had developed positively with regard to education, the economy, and technology.

However, in discussing these development indicators attention should be paid to the input, the process, the output, and the linkages of the entire system, rather than to the direct output of the sectors alone. Furthermore, planning tends to focus on the measurement of a tangible outcome, but there are many aspects of development that cannot be captured in this way, notably the social and human aspects. All the national development plans neglected these aspects.

At the initial stage of national development, agriculture was emphasized with a view to meeting both domestic and export needs. As early as the sixteenth century, the export of agricultural commodities was the result of foreign influences, which changed the economic and

169

production structures of the country. The increased demands of the external market expanded the area under cultivation. But this extensive growth resulted in a great loss of forest resources (table 1). One should note that in 1985 the majority of the population was still living in rural areas; the increase in production in the country occurred essentially through the exploitation of traditional technologies.

While primary and secondary education disseminated basic science and technology from Western countries, the need to combat foreign domination and the realization by national leaders that a static knowledge of S&T could no longer help the country adapt itself to a changing world opened the country to innovative ideas and concepts. Students were sent abroad for training in order to facilitate the introduction of innovations and accelerate the country's modernization. Yet, though the country was seriously pursuing innovative technology, little attention was given to the development of mechanisms to select, control, and adapt the imported items to match the country's resources.

The first National Economic Development Plan (1961–1966) focused primarily on developing agriculture to meet world market demands. The import substitution industry was also highlighted. During this transformation period, the government helped to provide the necessary infrastructure and to develop technical skills, and the private sector was urged to participate in production under the close guidance of the government.

The government introduced science and technology in two ways – first, by sending students abroad, and second by the purchase of technology goods. These two channels helped accelerate the acquisition of a technological capacity, yet it created a social cleavage: those in the urban sector benefited through educational opportunities and the utilization of imported technologies, while those in the rural sector had less opportunity to do so. Furthermore, the items acquired for industrial development were used mainly to produce goods for the local market.

Replicating or buying appropriate technology was not considered. In agriculture, though foreign technologies had some influence, the majority of farmers still used indigenous technologies. However, the output of agricultural products increased satisfactorily as a result of extensive cultivation.

In the second National Economic and Social Development Plan (1967–1971), the basic roles of government and private sector remained unchanged. The government continued to construct physical infrastructures, such as roads, railways, and irrigation dams, as well as

170

providing the rural community with important health services. The private sector, on the other hand, was being continuously urged to put more effort into the production of industrial goods. The government continued sending students abroad, and the purchase of technological items continued. There was a continued neglect of mechanisms for selecting and controlling foreign technology.

Within the agricultural sector, an increased use of modern production technologies, in the form of chemical fertilizers, pesticides, and small farm machinery, was pursued. Most of these, however, were imported. Although agricultural production increased tremendously, it did not keep pace with the increased production costs.

As a consequence of the second National Plan, certain undesirable phenomena emerged. These included a higher unemployment rate, a higher migration rate, and water pollution resulting from the drainage into waterways of chemical residues and waste materials from manufacturing. The government responded in the third National Economic and Social Development Plan (1972–1976) by the imposition of regulations and codes. Other measures were the expansion of compulsory education to neglected rural areas and an improvement in the quality of, and opportunities for, higher education. It was expected that the demand for higher technical skills would increase. Local physical structures, such as roads and local health care and rural development projects, were also emphasized during the third Plan.

Because of the package of policy measures adopted by the state during this period, industrial production was increasing at a high rate. Many of these products, particularly textiles, were mainly for local consumption. However, the industrialization of Thailand still had a number of barriers to breach.

The first of these was the continuous import both of foreign technologies for local manufacture and of materials – particularly iron-based materials – for industrial products (table 2). This led not only to a serious trade deficit, but also to a reliance on foreign support for industrial development. The government increased the number of S&T degree-holders, but most of these were mainly engaged in industrial management, process operation and maintenance, and product control sections. Another problem was the lack of selection in technology, which denied technologists the chance to improve their capabilities in order to progress to the replication and innovation stages of technological development.

During the 15-year period 1966–1980, rice output increased by 19 per cent, but the area under rice cultivation increased by 47 per cent.

Table 2. Expenses for imported steel and steel-based products (millions of baht)

Year	Non-electrical items for industry	Machinery and parts for agriculture	Tractors	Iron/steel	Other metals
1957	567	12	54	467	86
1962	1,232	19	133	479	147
1967	2,875	33	655	1,231	422
1972	4,706	36	345	2,046	1,043
1977	10,424	106	2,062	6,352	3,454
1982	19,329	164	1,679	11,323	5,811
1984	32,979	192	1,821	14,035	7,339

Similarly, during the six-year period 1974–1979, the gross amount of maize produced increased by 50 per cent, while the cultivated area increased by 61 per cent. This undesirable trend occurred at a time when the government was promoting the extensive use of modern production technologies, such as chemical fertilizers, pesticides, improved seed varieties, and improved techniques. The more the government emphasized the use of such technologies, the higher the cost of production became for the farmers.

By the fourth National Economic and Social Development Plan (1977–1981), industry was able to produce enough to meet domestic consumption needs. The government had invested considerably in the construction of the basic physical infrastructure for future industrialization. In a policy shift, it now established a policy of exporting industrial products. This also implied a shift of emphasis from agricultural exports to the industrial sector.

The policy, which gave effective economic incentives to entrepreneurs, was successful in terms of higher GDP rates. Yet the government had no concrete policy for developing technology on a self-reliant basis. The country continued importing foreign hardware technologies and iron-based materials for industrial purposes, increasing the trade deficit. S&T-trained manpower was still engaged primarily in machine operation and maintenance. But for the government, technology screening was not important as long as the country benefited from the exported products. And in the agricultural sector, although production rose the problem of the high cost of production was not addressed and farmers suffered.

During the fifth National Economic and Social Development Plan (1982–1986), the government continued its policy of industrial promotion for exports. This policy was reinforced by the discovery of petroleum. The policy for agricultural development also remained the same as in the preceding Plan. Experience with the Plan indicates that the poor structure of S&T development had not been sufficiently remedied.

Since the first National Economic Development Plan initiated in 1961, the country has followed a constant policy of purchasing foreign technologies, particularly hard industrial technologies and iron-based materials. The agriculture sector, in contrast, has been able to generate its own indigenous techniques for agriculture. But some modern production inputs in agriculture have been imported on a continuous basis. These indicate not only a heavy trade deficit, but also a lack of interest in developing one's own technology.

An evaluation of Thailand's present S&T situation: A macro-level study

The question of whether Thailand can achieve self-reliance in science and technology will be discussed by means of a techno-system model.

The general techno-system model shown in figure 2 has three major components, namely, the input, the techno-system, and the output. The input component is made up of a number of systems, functional subcomponents such as the infrastructure, manpower, and management. The techno-system component consists of subcomponents like the R&D system, the diffusion system, and the knowledge stock system. The techno-system used has both a direct and an indirect output.

Unlike the direct output, which is measurable, the indirect output seems to be complicated, requiring a knowledge of several disciplines – economic, social, and environmental – and resources to measure it. The economic impacts are primarily concerned with, for example, the proportion of the R&D cost to the product cost, and the proportion of the trade deficit to the production cost. The social impacts are concerned with the social capability of the country to absorb the relevant technologies being developed or introduced. The environmental impacts reflect pollution problems, particularly in terms of the capability of the country to control them. Finally, the resource impacts focus upon the capability of the country to control and utilize local resources.

There are seven possible levels of ability to use technology. Table 3 summarizes the level of technology use and the desirable education of users.

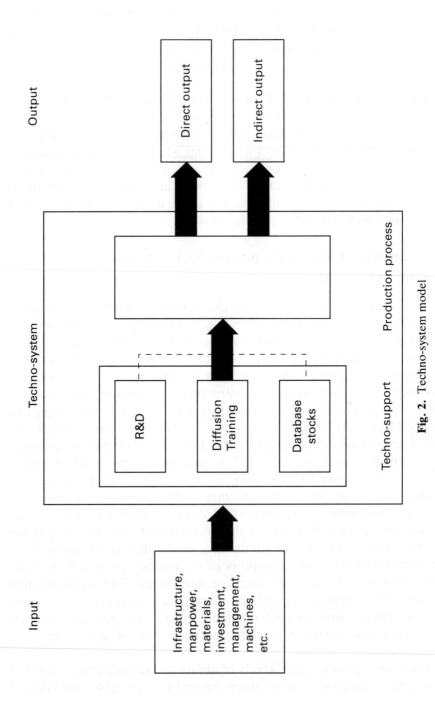

Fig. 2. Techno-system model

Table 3. Level of technology use and desirable education of users

Level	Description	Desirable education
1. Benefit from use only	User does not have to concern himself with the operation of technology	Any level
2. Copy/operate/maintain	User can copy or operate, implement or maintain	Primary
3. Judiciously select	User can make decision to select appropriate technology for his own use	Secondary
4. Replicate	User can copy but able to produce a better output	Vocational or engineering degree
5. Adapt/modify	User can adapt/modify technology to suit his own working conditions	Advanced degree in engineering or long experience
6. Innovate	User able to come up with new product using same concept	Advanced degree with research experience
7. Create	New product based on new concept	As (6) above

In Thailand, 91 per cent of the population has primary education, 8 per cent secondary, and 1 per cent university or vocational. The optimum level of technology use would involve the ability to select technology appropriate to given working conditions. For Thailand to achieve self-reliance and the ability to select technology, she must also be able to implement, manage, operate, and maintain the system.

Five concepts are used to describe the capability of the country to develop the technology suitable to its physical, economic, social, and cultural environments, which constitutes the essence of self-reliance. These capabilities are those of selecting, implementing, managing, operating, and adapting (fig. 3).

The system characteristics are primarily qualitative, namely feedback, diffusion, and memory. Each characteristic is further subdivided into variables which express further details. Each of these is given a value based upon the observations that appear in reports and individual interviews: 1 = absent or low; 2 = partly or medium; and 3 = present or high.

The result of the analysis of the techno-system in question appears in table 4. Knowledge stock indicators are:
1. Number of research institutions and laboratories.

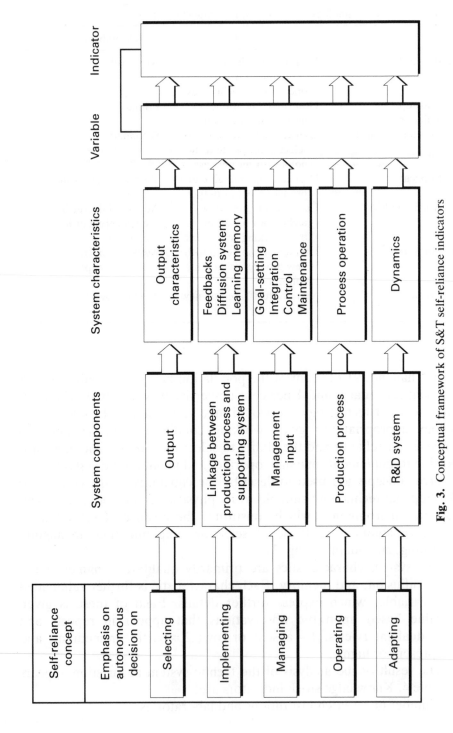

Fig. 3. Conceptual framework of S&T self-reliance indicators

Table 4. Indicators of Thailand's self-reliance in science and technology

System characteristic	Variables	Indicators	Empirical references	Value
Indicators of techno-system				
Goal-setting	Local autonomy	Foreign influence in policy formulation	A certain degree of foreign influence	2
		Existence of plan for local autonomy	– Little interference from government in the production process – Encouragement of more private participation in macro management	3
		Vertical integration	– Very weak vertical linkages at all levels	1
	Articulation of technosystem policies	Role of nation in policy formulation	– Policy formulation is mostly top-down	1
Control	Role of nation in corporate decision-making	Existence of corporate programme in setting up of national plan	Not clear	1
	S&T selection	Existence of S&T selection	No	1
	Output control	Existence of evaluation programme in the national plan	Not clear	1
	System control	Role of private sector in the system's operation	Some mention	2
		Existence of system evaluation in the national plan	Not clear	1
Dynamics	Adaptations and innovations in the relevant technologies	Number of adaptations and innovations in the relevant technologies	Very few innovations, some adaptations	1
		Quality of technological adaptation and innovation	Specific applications only	1

	Indicator	Measure	Description	Score
	Patterns of R&D outputs	Structural feature of R&D outputs	Mostly basic and some applied research	1
	Application of R&D output	Percentage of applicable R&D output to total	About 70% applicable with limited users	2
		Percentage of local R&D products to imports	Very low	1
System memory	Change in manpower with relative know-how	Rate of change of amount of S&T manpower	Increasing at high rate during last decade	3
	Documentation	Existence of local technical library	Subregional distribution	2
	Quantity and quality of document	Existence of historical technical statistics	Very few	1
System feedbacks	Linkage of process with local R&D	Support of local R&D by the production process	About 50%	2
	Linkage of production process with training and education programme	Utilization of locally trained technicians and engineers	Yes	3
		Utilization of local resource persons	Some foreign influences in policy formulation only	3
System maintenance	Adequacy of local technological educational system	Relevance of curriculum to country's development needs	Fairly independent, without proper linkage to country's development	2
		Quantity of graduates	Sufficient for local use	3
		Quality of graduates	Thai standard	2
	Adequacy of local supply of industry hardware and maintenance	Local supply of hardware	High percentage of imports	2
Interdependence/ integration	Network of the system	Existence of various components of the techno-system	Yes, but the performance is not impressive	2
		Interdependence/linkage of the subsystem	Rather poor	1

Indicators of outputs

Direct output	Quantity of product	Adequacy for local consumption	Weighted average of $\dfrac{\text{production}}{\text{consumption}} = 1$	3
		Percentage of export	Weighted average of $\dfrac{\text{export}}{\text{production}} = 0.012 - 0.37$	2
	Quality of product	Standard	Local standard controls (comparable to ISO)	2
Indicator output	Economic impacts	Individual B/C ratio	Agricultural production = 1; industrial production = ?	2
		National B/C ratio	?	?
		Percentage of S/D cost to production cost	Agricultural production = 10.05%; others = 87.8% (1980)	2
		Percentage of trade deficit to production cost	−5.5% (1980)	1
	Social impacts	Social capability in absorbing the relevant technologies	Operative capability only	1
		Unemployment	Increasing rate, partly due to improper use of S&T	2
	Environmental impacts	Pollution	Pollution in some main rivers and estuaries in the Gulf of Thailand	2
			Pollution in the food chain	
		Pollution-control capability	Industrial law but rather loose control in many ways	1
			No deficit control in the agricultural sector	
	Resource impacts	Local resource utilization	Use of local human resources	2
			Use of most available resources	
		Resource control capability	No practical control of proper resource utilization	1
		Output indicator		1.75
		Techno-system indicator		1.74
		Self-reliance indicator		1.745

2. Number of higher education institutions engaged in development of S&T.
3. Number of S&T degree-holders.
4. Number of scientific journals.
5. R&D expenditure.
6. Foreign technological collaboration.
7. Payment for imported technologies.

The analysis gives the self-reliance value of the techno-system as 1.74, which is rather low.

Self-reliance is thought to be weak in the following areas, where the average score is 1.0:

1. The structure of S&T organization reflecting a top-down approach.
2. No mechanism to select imported ideas, concepts, and tangible technologies so that they fit the country's environment.
3. Inadequate efforts to evaluate the progress of both the R&D programmes and the National Development Plan.
4. Insufficient adaptation and innovation, both in quantity and quality.
5. Low proportion of local technologies to imported ones.
6. Doubt about official statistics on R&D development.
7. Lack of linkages in the R&D subsystems.

Case-studies in agriculture

Eight micro-level case-studies were carried out in the agriculture sector to examine the self-reliance strategies. These were on crops (rice, sugar cane, and pineapple), livestock (dairy, cattle, pigs, and poultry), and fisheries (fish and shrimp). Details from the case-studies indicate that the use of technology is uneven, except in a remote farming community. Farmers in all parts of the country are now practising commercial rather than subsistence agriculture. Thai farmers' behaviour, it was found, is a rational response to social and economic conditions. Poor farmers are unwilling to cooperate in technology-oriented development programmes because they correctly perceive that the programmes often introduce uncertainties, largely because of the higher costs of production. However, once they perceive that assimilation of modern technology is beneficial, they adopt it. The use of technology for agricultural production is, however, uneven because of the technology costs as well as of farm prices. All case-studies at the micro level indicate that, in order to respond to an unstable environment, farmers practise a combination of traditional and modern technologies.

The commodities studied varied in commercial complexity. Rice, pineapple, and sugar cane serve as inputs into the agro-industry of the country. Rice is a primary product that is further processed in a rice mill, while pineapple and sugar cane are processed in a cannery and refinery respectively. Because of the small scale of dairy cattle enterprises, milk produced daily is supplied to small-scale processing plants operated by either the government or the private sector. Inland fish and shrimp production is also small compared to offshore fishery enterprises, and basically supplies the demand of local consumers. Keeping chicken and pigs, on the other hand, because of urban demands, can be a large-scale operation. Even here, however, operations are smaller in the countryside, and largely supply a slaughterhouse that supplies local market demand.

The analysis revealed moderate self-reliance values for rice, chicken, pig, dairy cattle, and shrimp and fish, and low values for pineapple and sugar cane (table 5). The values in all cases do not differ significantly, mainly because a large portion of the techno-system is controlled by a government-supported research organization. It is usually found that the goal settings for all research programmes are mostly determined by the government. Consequently, the number of projects carried out annually and of innovations in agricultural production, as well as the allocation of resources and the system's maintenance, is substantially controllable. A high self-reliance figure is therefore found in all cases for these characteristics. As a result of their freedom from control by the government sector and of other relevant factors, characteristics such as system feedbacks, interdependence, and part of systems maintenance have brought the self-reliance figure down.

The findings suggest a number of measures to remedy the weaknesses of the techno-system. These all concern the effective transfer of technology. The case-studies indicate that there is insufficient evidence to support the proposition that the more sophisticated the technology, the greater its application. Rather, the reverse effect is observed.

These micro-level case-studies have led the study team to put forward two propositions.

Proposition 1. Neither the level of modern S&T nor the extent of its use correlate with self-reliance. The appropriateness of the S&T level, however, correlates significantly with the user's ability.

Proposition 2. There is little correlation between the techno-system value and the techno-system output value. The real meaning of self-reliance in S&T is therefore a functional combination of these two values in accordance with the appropriateness of the S&T and the user's ability.

Table 5. The micro-level stage of self-reliance

	Rice	Pineapple	Sugar cane	Poultry	Pigs	Dairy	Inland shrimp/fish
Linkage with industry	Rice mill	Cannery	Refinery	Slaughterhouse	Slaughterhouse	Cooperative milk-processing plant	Market
S&T self-reliance value	2.16	1.80	1.67	2.15 (situational)	2.04 (situational)	2.17	2.00
Level of technology use[a]	**	****	****	***	***	****	****
Strengths of the system	R&D dynamic Goal-setting Control Maintenance	As for rice	As for rice	As for rice			
Weaknesses of the system	Linkages between production process and supporting system	As for rice	As for rice	As for rice			
Improvement suggested	Appropriate R&D output Training Publication Market information Management skill	As for rice	As for rice	As for rice			

a. * = very traditional; ***** = very modern. Number of asterisks in one column reflects position between these extremes.

A desirable path

Thailand has attempted to develop technology as a springboard to industrialization. However, one could map out a more desirable path, given the country's limited resources.

Two different concepts, "supply push" and "demand pull," help decide the investment of resources in technology. The former is more applicable to a developed country where unlimited resources exist to be invested in a desired technology. In a resource-poor country like Thailand, demand pull is more relevant, since the limitation in national endowments has to be kept in mind so as not to direct resources to a low-priority sector.

We have used six significant factors to decide whether and where the country should invest: physical infrastructure, natural resources, S&T resources, economics, cultural heritage, and manpower resources and the S&T resource base.

Thailand is considered ready to pursue a path of industrialization and has invested considerable resources in building an infrastructure for industry. Those infrastructures available even at the village community level include, for example, transportation and electrical power. In contrast, the resources invested in agricultural infrastructure are inadequate: irrigation is one example. With limited national resources, the country chose to invest more in industry than in agriculture.

Since the first National Economic Development Plan, Thailand had continuously imported iron material and other industrial items, causing a trade imbalance in the country. This indicates that the heavy investment in industrial infrastructure and the government's incentives have not developed a strong industrial base for the country (table 4). The industrial sector is still at the operative stage of mastery of software technology. In agriculture, on the other hand, software capability (i.e. the farming system) has reached the replicative stage, although capability in hardware technology (i.e. farm machines) is at the operative stage. Because of its modest resource demands, Thailand should promote more development in the agricultural sector.

The rainfall pattern and the type of arable land indicate that all regions of the country except the North-east are suitable for general agricultural production. In some parts of the North-east, rainfall is sufficient for vegetable growing. By and large, Thailand's natural endowments favour agricultural production.

Proportionately, Thailand possesses more land for rice production than the Philippines, Japan, and the Republic of Korea. However, its

183

yield per unit area is lower than that of all these countries, as well as of India and China (table 6). Government statistics also show that the longer the country continues its present production practices in agriculture, the less the productivity of the crops will be. Statistics indicate that increased crop production has in fact been obtained from the expansion of land under cultivation. A favourable policy for agricultural development, economic incentives, and suitable agricultural technologies at the farm level could change this.

In the Republic of Korea, the private sector has to share the cost burden in S&T development. A law was enacted to encourage, indeed force, the industrial community to train technicians by, for example, providing vocational schools. Since 1982, the Korean government has devised a mechanism to encourage the private sector to increase investment in R&D and in related technological activities. As a consequence, investment by the private sector in technological capability, including R&D, has increased sharply. A similar situation exists in Japan. Much constructive experience in technological self-reliance can be drawn from the Japanese and Korean examples. These experiences could be fruitfully used to direct S&T development in Thailand.

To increase "hardware" capability in Thailand, it is suggested that:

1. All involved in "hardware" work (e.g. skilled workers and producers), regardless of the size of the factory, should be registered.
2. Skill development should be promoted through training programmes.
3. Academic institutions should be encouraged to achieve higher standards.
4. Professional associations should be established as centres of exchange of perspectives, skills, and production. Central institutions with responsibility for developing and replicating skills and production should also be established, and a network for communicating technological information among institutions should be set up to facilitate technology transfer.

To increase self-reliance, efforts should be made to carry out exploration for raw materials that are now imported as well as to develop alternative materials. R&D in agriculture and industry should ensure that research results are more accessible to users. All research programmes, moreover, should be continuously evaluated.

The study results indicate that the development of technical skills through an exchange training programme would be useful. Materials science, metal processing, and industrial process design are priority disciplines. An international network in S&T information, aimed at achieving self-reliance, should be another focus of cooperation.

Table 6. Relationship between the harvested area of rice and the yield per unit area

Country	Harvested area (millions of rai)[a]					Yield per rai (kg)				
	1977	1978	1979	1980	1981	1977	1978	1979	1980	1981
Thailand	55.68	55.84	54.08	56.86	56.90	225	312	291	305	312
India	251.76	243.62	243.58	253.12	250.00	314	331	261	328	328
Philippines	21.93	21.68	21.11	21.56	21.85	314	332	345	345	353
China	231.73	210.95	216.19	213.63	215.50	559	655	680	666	678
USA	5.68	7.51	7.25	8.33	9.61	791	804	825	790	874
Japan	17.23	15.92	15.60	14.85	14.23	987	1,027	958	820	901
Republic of Korea	7.55	7.68	7.70	7.62	7.65	1,098	1,086	1,023	787	919

a. 1 rai = 0.16 hectare.

185

Human resources development for industry requires that the majority of the population possess a basic knowledge of science and technology. Education is investment capital. Yet the people in Thailand are holding back progress, since the majority do not possess sufficient technical skills and knowledge for national development, particularly in industry. Formal, informal, and non-formal education could be used to change this situation. Higher education opportunities for the rural population should be rapidly expanded.

Thailand would benefit from an educational exchange programme, which would increase the capability of its human resources. Mutual assistance in technology should be developed through training programmes. A country cannot develop in a vacuum – it needs to acquire the relevant essential information. Information from indigenous or exogenous sources has to be organized in networks, so that it can be widely accessed by institutions and used in an open-minded and cooperative manner.

Bibliography

Aye, Thaung. "Impact of Agricultural Mechanization in Burma." Paper submitted to Seminar on Agricultural Mechanization in Developing Countries, UIATC, Uchihara, Japan, 1980.

Chulalongkorn University. *Proceedings of a Seminar on People Participated Development: Strategies*. Bangkok, 1984. (In Thai.)

Department of Agriculture. Ministry of Agriculture and Cooperatives. *Proceedings of the Workship on Grain Post-harvest Technology*. Bangkok, 1978.

Economic and Social Commission for Asia and the Pacific. "Study on Human Resources Development: Its Technological Dimensions." Draft report submitted to Expert Group Meeting on Human Resources Development: Its Technological Dimension II, ESCAP Building, Bangkok, December 1985.

Embree, F. John. "Thailand: A Loosely Structured Social System." *American Anthropologist* 52 (1950), no. 2: 181–193.

Faculty of Engineering, Chulalongkorn University. *Proceedings of Second National Conference on Self-reliance in S&T for National Development*. Bangkok, 1986.

Farm Machinery Industrial Research Co. *AMA Magazine for Agricultural Mechanization in Asia, Africa and Latin America* (Tokyo): 12 (1981), nos. 1, 2; 14 (1983), no 1; 15 (1984), no. 4.

Gifford, R.C. *Agricultural Mechanization Strategy Guidelines for Thailand*. Bangkok, 1981. (UNDP/FAQ project no. THA/79/005.)

Group Business Co. Ltd. "A Study on the Iron and Steel Casting Market in Thailand." Final report submitted to the Construction Material Marketing Co. Ltd, Bangkok, 1977.

Hathway, Gordon. *Low-cost Vehicles*. London: Intermediate Technology Publication, 1985.

IMC. *Policy and Program for the Promotion of Small Scale and Regional Industries.* Industrial Restructuring Study. Bangkok: NESDB, 1984.

———. "Industrial Restructuring in Machinery Industry." Report presented to UNDP/UNIDO-NESDB, Bangkok, 1986.

———. *Technology Development and Promotion for the Engineering Industries.* Industrial Restructuring Study. Bangkok: NESDB, 1984.

Indian Institute of Management. *Proceedings of an International Workship on Rural Poor: Their Hopes and Aspirations.* Jamua, India, 1981.

Kasetsart University Research and Development Institute. "The Consequences of Small Rice Farm Mechanization in Thailand." Workshop paper. Bangkok, 1983.

McClelland, David C. *The Achieving Society.* Princeton, N.J.: Van Nostrand, 1961.

McClelland, David C., and David G. Winter. *Motivating Economic Achievement.* New York: The Free Press, 1969.

Ministry of Science, Technology, and Energy. *Proceedings of the First Science Congress on National Social and Economic Development with Science and Technology.* 1984. (In Thai.)

Mulder, J. A. Niels. "Origin, Development, and Use of the Concept of 'Loose Structure' in the Literature about Thailand: An Evaluation." In: Hans Dieter Evers, *Loosely Structured Social System in Comparative Perspectives.* New Haven, Conn.: Yale University Press, 1969.

NESDE. *Proceedings of a Seminar on Research Findings on Leading Branch of Industry for Job Creation in Sixth National Economic and Social Development Plan.* Pattaya, 1985. (In Thai.)

Office of Agricultural Statistics, Ministry of Agriculture and Cooperatives. *Agricultural Statistics of Thailand: Crop Year 1981/82, 1983/84 and 1984/85.* Bangkok, 1982–1986.

Posada, R. "Self-reliance in S&T for National Development: The Philippines as a Case Study." Interim report presented at an International Workshop on Self-reliance in S&T, Beijing, 1985.

Prempridi, Thamrong. "Low-cost Transport in Thailand: Case Study of Chiang Mai and Chiang Rai." Report submitted to ILO, Geneva, 1986.

Prempridi, Thamrong, Mongkol Dandhanin, and Vishan Poopath. "Self-reliance in S&T for National Development: Thailand as a Case Study." Final report submitted to UNU, Tokyo, 1985.

Regional Network of Appropriate Technology for Rural Development in SE Asia and Pacific. *Final Report and Proceedings of the Seminar on Farm Machinery and Rural Industry.* Bangkok: Faculty of Engineering, Chulalongkorn University, 1984.

Research and Data Resources Co., Ltd. "The Feasibility of the Project to Develop Engineering Industries in Thailand." Report submitted to Department of Industrial Promotion, Bangkok, 1981.

Said, Zaidir. "Impact of Farm Mechanization in Indonesia." Paper submitted to a seminar on Agricultural Mechanization on Developing Countries, UIATC, Uchihara, Japan, 1980.

Shishido, Toshio, Takeshi Hayashi, and Ichiro Inukai. "Technological Self-reliance of Japanese Food Processing Industry." Report submitted to UNU, Tokyo, 1985.

Somtrakoon, Kla. *A Demonstration Project Proposal on Intensive (Self-sufficiency) Farming and Appropriate Technology.* Bangkok: Ministry of Education, 1983.

Technical Committee of Agricultural Machinery Production. "Agricultural Machinery

Production Project." Document for the Fourth Technical Committee Meeting, June 1984, Kampangsaen, Thailand. (THA/79/005)

Thailand National Commission for UNESCO. *Bulletin of TNC for UNESCO*, vols. 10–17. Bangkok: Ministry of Education, 1978–1985. (In Thai.)

Toet, A.J. "Assessment of Existing Farm Mechanization in relation to Prevailing Farming Systems." Report for FAO/UNDP project no. THA/79/005. Bangkok, 1983.

Tsutumu, I.S.A., *Studies on the Conventional Farming Tools and the Evolution of Farming Systems in Southeast Asia*. Mie, Japan: Faculty of Agriculture, MIE University, 1983.

Uchihara, Yasuo M. "Agricultural Mechanization in Developing Countries." Summary of seminar, International Agricultural Training Center, Tokyo, 1980.

UNIDO. "Engineering Industries in Thailand." Seminar on industrial restructuring organized by RESCOM Secretariat and UNIDO, NESDB, and IMC, Bangkok, 1984.

———. *Proceedings of a Regional Seminar on Engineering Industries in Thailand*. Bangkok, 1985.

Unisearch, Chulalongkorn University. "A Proposal to Draw Up a Master Plan for the Development of Farm Machinery in Thailand." Technical document submitted along with tender documents to Ministry of Industry, Bangkok, 1986.

Vacharotayan, Sorasith. *Proceedings of the International Seminar on Environmental Factors in Agricultural Production*. Bangkok: Kasetsat University, 1985.

Van Dijck, Pitou, and H. Verbruggen, eds. "Export-oriented Industrialization and Employment: Policies and Responses with Special Reference to ASEAN Countries." Manila: Council of Asian Manpower Studies, 1984.

Wick, J., and S. Buengsung. "In-depth Study of Mechanization Needs in San Patong District, Chiang Mai Province." Agricultural Machinery Production Project. Bangkok, 1984. (UNDP/FAO project no. THA/79/005.)

Williams, J. F. Report on the Manufacture of Farm Machinery in Thailand. Agricultural Machinery Production Project. Bangkok, 1984. (UNDP/FAO project no. THA/79/005.)

World Bank. *World Development Report 1984*. New York: Oxford University Press, 1984.

5

The Philippines

Alvin Toffler's paradigm of the three waves of civilizations, following and sometimes overlapping one another, represents the major stages of S&T development.[1] His civilizations also reflect an ascending level of scientific and technological sophistication. They are characterized by the following technologies:

1. First-wave technologies, comprising the pre-industrial technologies which are labour-intensive, small-scale, decentralized, and based on empirical rather than scientific knowledge. The intermediate, appropriate, or alternative technologies based on the Schumacherian philosophy of "small is beautiful" also fall into this category.

2. Second-wave technologies, comprising the industrial technologies that were developed between the time of the Industrial Revolution and the end of the Second World War. These are essentially based on the principles of classical physics, classical chemistry, and classical biology.

3. Third-wave technologies, comprising the post-industrial or high technologies which are called science-intensive because they are based on our modern scientific knowledge of the structures, properties, and interactions of molecules, atoms, and nuclei. Among the important high technologies are micro-electronics, robotics, computers, laser technology, optoelectronics and fibre optics, genetic engineering, photovoltaics, polymers, and other synthetic materials. Some of the representative types of technologies in the first-wave, second-wave and third-wave classes are tabulated in the S&T taxonomical matrix given in table 1.

Table 1. S&T taxonomical matrix

Type of technology	First-wave technologies	Second-wave technologies	Third-wave technologies
Materials technologies	Copper, bronze, iron, glass, ceramic, paper	Steel, aluminium, dyes, plastics, petrochemicals	Polymers, semiconductors, liquid crystals, superconductors
Equipment technologies	Plough, lathe, mills and pumps, spinning wheel	Engines, motors, turbines, machine tools	Laser tools, microprocessors, robots
Energy technologies	Wood and charcoal, wind power, water power	Coal, oil, hydroelectric power, geothermal power	Solar cells, synthetic fuels, nuclear fusion
Information technologies	Printing, books and letters, messengers	Typewriter, telephone, radio, telegraph, TV	Computers, fibre optics, artificial intelligence
Life technologies	Traditional agriculture, animal breeding, herbal	Mechanized agriculture, surgery, antibiotics, food	Hydroponics, artificial organs, genetic engineering

It is possible to define five discernible stages in the development of a national technological capability. These are, in ascending order: (1) operative capability; (2) adaptive capability; (3) replicative capability; (4) innovative capability, which is the ability to make significant modifications and improvements on the basic design of existing technology; and (5) creative capability, the ability to design and produce an entirely new and revolutionary technology.

The attainment of stage 5 (creative capability) represents technological mastery in a given country.

The notion of technology is complex, with numerous links to other complex notions: the conceptual framework that we use in this study, including the case-studies, is the techno-system shown in figure 1. This shows the ends and means of organized production as well as the growth and evolution of the useful stock of technical knowledge.

The stock of relevant knowledge in the information subsystem interacts with the other components through the flow of information and feedback processes. It consists not only of scientific and technical knowledge, but also managerial, banking, legal, and other skills.

One mechanism that stimulates the growth of the stock is research and development (R&D), a component of the techno-system linked

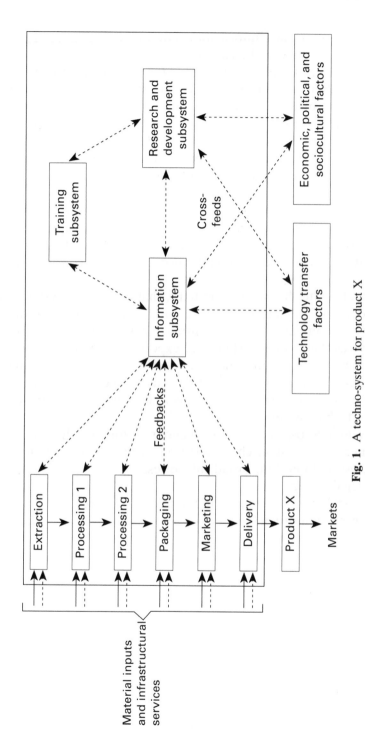

Fig. 1. A techno-system for product X

191

with the others through information flows and feedbacks. The R&D component is the source of changes.

The definition of inputs and outputs for the techno-system also defines essentially its system boundaries and structure. In the copper industry, for example, we could consider either copper metal or copper wire to be the principal output of the techno-system. The principal input could be either just energy or energy and copper concentrates. These choices imply various configurations of the system components. To reduce arbitrariness, the principal output is limited to consumer products or intermediate products. The inputs could either be endogenous (internal to the system) or exogenous (outside the system). Of the various system components, material inputs, capital, and unskilled labour are defined as exogenous, while skilled labour and managerial inputs, which are essentially information, are considered endogenous.

One could state, by way of summary, that the techno-system is conceived to be an organized structure for the creation of products to satisfy a set of human needs. Its central feature is the knowledge stock which acts as the source of skills and expertise in the operation of the various components. It provides the mechanism for systems' memory and learning. Technology refers to the knowledge and skills, either in software or embodied in hardware, associated with productive components of a techno-system.

The system "crossfeeds" are exogenous factors which greatly affect (i.e. influence the system characteristics of) the techno-system. These may be classified into four broad categories.

Political/legal factors:
- Political stability/government and political structures.
- State perception of S&T.
- State priorities in S&T.
- State incentives, disincentives.
- Endogenization of S&T.
- State policies on technology transfer.
- Policies of major trading partners.
- Activism of engineers and scientists.
- Interests of political leaders.
- Capacity for policy implementation.
- Consensus on development goals.
- Acceptance of meritocracy.
- Existence of policy instruments favouring self-reliance.
- Existence of vested interest for technological dependence.
- Corruption.

Sociocultural factors:
- History of S&T.
- S&T tradition.
- Commitment to self-reliance in S&T.
- Social environment for successful technology transfer.
- National pride.
- Social equity in technology development.
- Existence of a "techno-class."
- Educational levels in S&T.
- National S&T potentials.
- Existing technological capacity.
- Social cohesiveness and stability.
- Self-reliant attitudes of scientists/engineers.
- Class character of technology.
- Attitudes favouring technological dependence.

Economic factors:
- Economic development philosophy.
- Existing structure of the national economy.
- Economic roles of the private and state sectors.
- Local market size.
- Economic dualism.
- Strong local demand for foreign products.

Technology transfer factors:
- Transfer mechanisms.
- Capabilities for technology choices.
- Learning effects of technology transfer.
- Costs of technology transfer.
- Terms of technology transfer.
- Characteristics of technology.

We use a two-level definition of S&T self-reliance. At the macro level, self-reliance is defined as at least the existence of replicative capacity in all types of second-wave technologies. These are the entries in the third column of table 1. This definition may be complemented by choosing some values of the indicators in table 2.

At the micro level, self-reliance is associated with a specific integrated production system. It must be expressible in terms of systemic characteristics such as goal-setting, inputs and outputs, dynamics, control, learning and memory, etc. This is in contrast to the macro definition of S&T self-reliance, which is a definitive state of a country's S&T capacity. For example, it is only meaningful to talk about self-reliance in copper wires, or personal computers, or refrigerators. The concept, therefore, is micro.

Table 2. Comparative education indicators

| | Number enrolled in primary school as percentage of age-group | | | | | | Number enrolled in secondary school as percentage of age-group | | Number enrolled in higher education as percentage of population aged 20–24 | |
| | Total | | Male | | Female | | | | | |
	1965	1983	1965	1983	1965	1983	1965	1983	1965	1983
India	74	85	89	100	57	68	27	34	5	9
China	89	104	–	116	–	93	24	35	–	1
Philippines	113	114	115	115	111	113	41	63	19	26
Thailand	78	99	82	101	74	97	14	29	2	22
Republic of Korea	101	103	103	104	99	102	35	89	6	24
Japan	100	100	100	100	100	100	82	94	13	30

Source: World Bank, *World Development Report*, 1986.

The historical roots of technological dependence

Even before their contact with Western cultures, Filipinos already had an alphabet, some mathematics, a calendar, and a system of weights and measures. They were engaged in rice farming, fishing, and the mining of gold. Medicine based on local herbs was practised. Small boats and ships up to 2,000 tons were being constructed out of logs.

The Spaniards introduced the manufacture of lime, cement, and bricks and the use of concrete materials. Primary education was started by the Spanish missionaries in 1565. There were about a thousand of these parish primary schools by the end of the sixteenth century. The Spaniards also started higher education in as early as 1597 with the establishment of the Colegio de Cebu (now the University of San Carlos), and the University of Santo Tomas opened in 1611. Admissions to these schools were limited to a select few.

The emphasis in the church schools was on classical learning, specifically Latin, Greek, philosophy, the humanities, and law. Although medicine and pharmacy were taught, the natural sciences and engineering were generally neglected. The educational system, primarily based on the propagation of Roman Catholicism, did not foster a scientific tradition of scholarship.[2] On the contrary, it reinforced the superstitious, pre-scientific outlook of the existing folk beliefs.

The teaching of science was disdained and Filipino students were discouraged from its pursuit. The emphasis was on rote learning. The objective of the lesson, for example, was not to teach physics, but to convince Filipino students that they were incapable of learning physics. Yet the Spanish system produced Filipinos whose liberal education was comparable to that of the graduates of European universities.

In the Spanish colonial period, the cultivation of sugar and coconuts was started, and to support these activities the first agricultural school was established in Manila in 1861. Since then, sugar and coconuts have become the prominent elements of the Philippine economy.

The significant change during the American colonial period (1898–1946) was the establishment of an alternative to sectarian education. A department of Public Instruction was created. American teachers were imported, and English was used as a medium of instruction. In 1901, a Bureau of Government Laboratories (now the National Institute of Science and Technology) was established and concerned itself initially with activities related to chemistry and tropical diseases. In 1908, the first state university, the University of the Philippines (UP) was established. In the following year, 1909, the College of Agriculture was set up in Los Banos. In 1910, the College of Medicine was organized from the already existing Philippine Medical School. In 1926, scientific research was started at the College of Veterinary Medicine, and the School of Hygiene and Public Health was added to the University of the Philippines.

It is interesting to note that, as in the Spanish period, the focus of the American period was also on agriculture and the medical sciences. Industrial technology was initially relegated to the vocational level at the Philippine School of Arts and Trades. This bias is also reflected in the emergence of scientific periodicals. *The Philippine Agricultural Review* was first published in 1908, whereas the UP *Natural and Applied Science Bulletin* was started 22 years later. Even today, there are no specialized journals in physics. The history of the formation of scientific societies also reflects this uneven development. The Philippine Medical Society was organized in 1901 while the Philippine Society of Civil Engineers was formed only in 1933. The early bias towards agriculture and medical sciences was also prominent in the manpower training programme.

The Philippines was effectively transformed into an exporter of agricultural products and raw materials and an importer of manufactured goods. This hindered the emergence of economic self-reliance and industrialization. There was practically no demand for research engineers and physical scientists. The emphasis was on agricultural and medical research.

The momentum of this colonial policy has continued up to the present. Caoili[3] points out that factors associated with this colonial condition resulted in the cultivation of Filipino tastes for American brands and products. Cultural imperialism also critically influenced the outlook of the nascent Filipino scientific community.

In 1934, the American colonial government sanctioned the formation of the National Research Council of the Philippines, which was patterned after American models. Filipino scientists and their research were more relevant to the American condition, since the US was where they obtained their training and where their peers resided. Beyond the social effects of colonialism, the impact on the industrialization process itself has been profound.

As Yoshihara points out,[4] the entrepreneurial class in the Philippines dramatizes its colonial origins. Only one-third of entrepreneurs today are native Filipinos, the other two-thirds being mostly foreigners. Even during the early years of independence, Philippine industries were dominated by the Americans.

After the end of direct American rule in 1946, the uneven development of S&T in the Philippines continued. Most of the scientific organizations established by the independent Philippine government were also predominantly agriculture-based. The physical sciences, engineering, and mathematics continued to be neglected.

In 1956, a National Science Board was established by Republic Act 1606 to promote scientific, engineering, and technological research. In the same year, the Chairman of the Senate Committee on Scientific Advancement submitted a "Report on the Status of Science in the Philippines" to the President. Among other things, it recommended "an all-out financial support of scientific work and the establishment of a coordinating agency to handle scientific matters." This gave birth to the Science Act of 1958 (Republic Act 1067), which abolished the newly established National Science Board and created the National Science Development Board (NSDB). As reflected in the expenditures for R&D, the emphasis continued to be on agriculture and medicine, which accounted for more than half of all R&D funds. Basic research in the physical sciences was given something like 1–3 per cent of the total R&D budget, and applied industrial research about 5–15 per cent. According to NSDB figures for the 1960s, there were more physical scientists and engineers engaged in R&D, together constituting about 68 per cent of the total R&D workers. Life scientists (including medical and agriculture) were only about 15 per cent of the total. Thus, R&D expenditures were also biased in favour of agriculture and medicine.

The year 1968 is significant in the history of S&T in the Philippines. Presidential Proclamation No. 376 provided NSDB with a 35.6 hectare area in Bicutan to house the future Bicutan Science Community, consisting of research laboratories, pilot plants, science museum, etc. Moreover, the Congress of the Philippines passed Republic Act 5448, which imposed new taxes for a Special Science Fund to finance scientific activities for the next five years.

In the early 1970s, NSDB's principal concern was the infrastructural development of the science community. Most of the Special Science Fund was used for construction of the buildings of the National Science and Technology Authority (NSTA) and the other institutions.

The gross national expenditure for S&T for the period 1970–1975 varied from 0.21 to about 0.48 per cent of the GNP. Almost one-half of the research grants went to the University of the Philippines (UP). The significant developments in this decade were the establishment of the Philippine Council for Agriculture and Resources Research

197

(PCARR) and the Technology Resource Centre (TRC). PCARR became the effective research coordinating mechanism in the agricultural sector, resulting in more efficiency in the allocation of resources. This further strengthened the already dominant role of agriculture. The creation of TRC outside the orbit of NSDB was only the beginning of the dismantling and weakening of NSDB's monolithic hold on Philippine S&T. In this period, the Metals Industry Research and Development Centre and the Philippine Textile Research Institute were transferred from the NSDB to the Ministry of Trade and Industry. The National Computer Centre was established under the Ministry of National Defense. The TRC operates a technobank and a computerized database connected to foreign and local databases. The NSDB was pre-empted by others in the new and vital information technologies.

In 1982, NSDB was reorganized into a National Science and Technology Authority (NSTA) with four sectoral councils patterned after PCARR. In spite of this, however, NSTA was outside the mainstream of the Philippine industrialization programme. The Ministry of Trade and Industry (MTI) was supervising the Technology Transfer Board and the establishment of the country's major industrial projects. On the other hand, the TRC was implementing the so-called Technology Utilization of Energy under the Philippine National Oil Company. The control of MTI and TRC was in the hands of non-scientists. The management of S&T development in the Philippines was fragmented among various agencies. In spite of the transformation of the NSDB into an NSTA, it has, in fact, been considerably weakened by the loss of control over some of the important elements of national S&T development.

S&T policy: Rhetoric and reality

During the American rule in the Philippines, science and technology policy was outwardly benevolent and paternalistic but implicitly colonial in purpose. The early articles in the first issues of the *Philippine Journal of Science* were mostly aimed at investigating the country's natural resources, which were extremely useful to the colonizers. There were inventories of flora and fauna in various places, analyses of local minerals, taxonomy, geologic explorations, and tropical medicine. There was practically no "frontier" research in which the motive was simply to discover and elaborate the laws of nature. The Filipino scientists generally served as apprentices to the American researchers.

The scientist as classifier, data-gatherer, and taxonomist became the model of the Filipino students. Many of the first American-trained Filipino scientists were cast in this image. The scientist as a technician rather than a discoverer became a tradition which persists up to the present time. This is the taxonomy tradition in science in the Philippines. Its non-biological equivalents are, for example, the irradiation of various local materials, the chemistry of Philippine natural products, and the geology of various sites in the country.

There was no serious attempt to introduce industrial technology in the Philippines. The first colleges were the College of Agriculture (1909) and the College of Medicine (1910). Industrial technology was introduced later, in 1926, with the establishment of the Philippine School of Arts and Trades. Although a College of Engineering was established at the UP in 1910, engineering was not actively promoted, with the exception of civil engineering, which was needed for the construction of public facilities and for the surveying and mapping of the country. It was perhaps for this reason that the Philippines today has significant capability in this field. Philippine construction and surveying firms are carrying out projects in other countries.

The recognition of the importance of science and technology in the Philippines has not been wanting. In the 1935 Constitution, there is a provision (Article XII, section 4) which affirms: "the State shall promote scientific research and inventions." In the 1973 Constitution, the provision even clarified the role of S&T in development. Article XV, section 9(1), states: "The State shall promote scientific research and invention. The advancement of S&T shall have priority in national development."

The first concrete step in the implementation of the constitutional policy on science and technology was the enactment of the Science Act of 1958 (Republic Act 2067). It reaffirmed, in no uncertain terms, the belief in S&T as a tool for national development.

However, the picture that emerges from the actual allocation of government resources belies these good intentions. Upon its creation, the National Science Development Board was given a lump appropriation for five years. During this period, the growth rate in R&D expenditure was apparently a healthy 12.3 per cent. However, the average growth rates of all other government expenditure at that time was 13.0 per cent. Thus, the S&T sector received no special treatment. On the contrary, after the lump sum period, when NSDB had to compete with other government agencies, the growth rate for R&D funds was less than 1 per cent.[5]

In more recent times, the appropriation for NSTA has been, in general, declining as a percentage of the national budget. This is shown in figure 2. The sudden increase in 1983 was merely due to the reorganization of NSDB into NSTA: some new agencies were created or were attached to NSTA, resulting in an apparent increase in appropriations. Figure 3, which shows the R&D budget of NSTA as a fraction of the national budget, demonstrates the government's real attitude towards S&T research.

In sharp contrast with the financial reality, the rhetoric of presidential statements of policy have been bold, ebullient, and encouraging. Support for science in the Philippines has become something like motherhood statements. In the budget message of 1957, the President said: "Scientific research shall be intensified and accelerated, scientists adequately paid, because the results of our scientific investigations are the bases of economic and social progress." In 1966, in the State of the Nation Address, the President stated: "Our best efforts shall be directed to encourage the development of research." In 1968, the President once again bravely asserted that the country would "put scientific research on a systematic and continuing basis and it [in reference to a proposed, but never realized, science centre] will become the focal point of the scientific effort of the nation."

These statements were by three different Presidents of the Republic. Statements of this nature are common during science weeks, sponsored by the NSTA. NSTA usually takes them seriously as policy statements and formulates programmes accordingly, only to be disappointed at the actual results.

There have been three major policy episodes in science and technology in the Philippines, corresponding roughly to the present, the 1970s, and the 1960s. In the 1960s, the slogan was "import substitution," which mimicked the economic thrust of the government. Four priority areas were identified:
1. Basic needs and import substitution.
2. Quality improvement of exports.
3. Waste materials and product utilization.
4. Science promotion and education.

Looking back and evaluating the tangible results of this effort, it was only in education that the policy possibly made some mark. This was simply because of the continuous support of NSDB scholarships. Some curricular materials for science subjects were developed at UP.

In the 1970s, the battle cry was "mission-oriented research." Although several S&T missions were identified and planned, none was

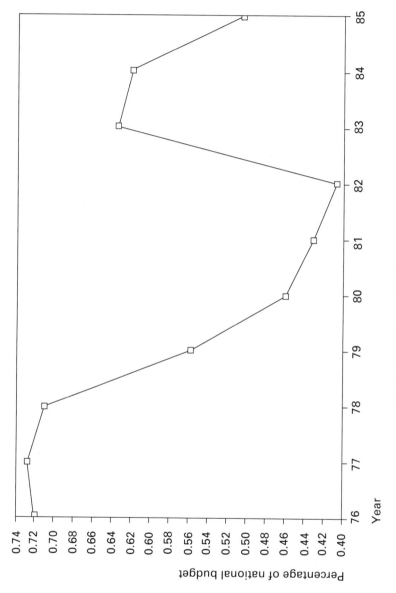

Fig. 2. NSTA general budget, 1976–1985

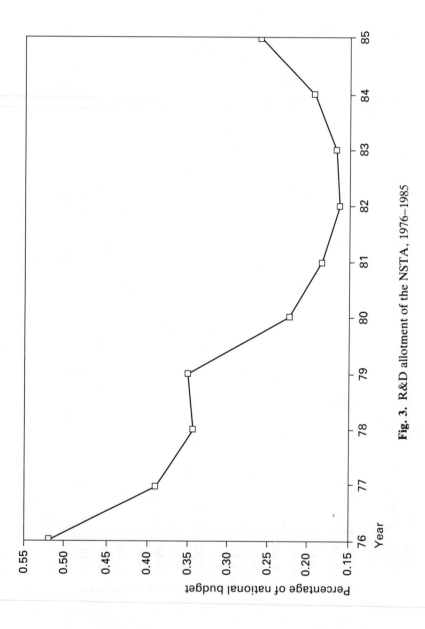

Fig. 3. R&D allotment of the NSTA, 1976–1985

actually funded or implemented. The inertia of the R&D community in doing what they have been doing, and the antipathy toward the Marcos administration, ultimately led to the demise of the policy and the NSDB administration that supported it. The policy package also included organizational reforms of the NSDB. These were the only ones that were actually realized. Organizational changes are always easier to undertake than programmatic changes.

The general policies of the NSTA for the 1980s were the following:

1. The strengthening of the support system for industries, and the emphasis on mission-oriented research for agriculture, natural resources, health, and energy.
2. The provision of increased resources for S&T, particularly for R&D and manpower development.
3. The strengthening of regional S&T institutions in order to encourage industrial dispersal and country development.
4. Encouraging the scientific community to evolve largely self-regulatory but centrally coordinated R&D organizations that would promote creativity and minimize administrative impediments.

"Demand-pull strategy" was the policy strategy of the early 1980s. This was accompanied by an elaborate National Science and Technology Plan. To a certain extent, the demand-pull strategy accomplished its goals. Some consumer products, such as soy sauce, bath soap, salt, and charcoal, were produced in cooperation with private industry and eventually marketed.

The National Science and Technology Plan was obviously the work of a committee. It contained a good number of priority areas, from ecology to micro-electronics, that could not possibly be supported at any reasonable level by the meagre funds of the NSTA. As previously, no specific missions were identified and pursued under a mission-oriented approach. The flamboyant commitment made by statutes and plans to S&T had not been matched by the outlay of financial resources.

An examination of the kind of research projects undertaken by the NIST from 1946 to 1982 shows a continuation of the emphasis on the analysis and use of Philippine natural products: earthenware from Philippine clays, sweet potato flour, powdered dilis (dried fish), oxalic and oxalates from Philippine vegetables, cottonization of ramie fibers, and others of the same genre.[6]

Thus, there was no essential departure from the colonial S&T policy. The only difference was that there were no American overseers. This kind of research has acquired so much momentum that it continues to command the commitment of a significant fraction of R&D funds. While its importance cannot be denied, the resulting inhibition of the development of other scientific fields is unfortunate. The colonial policy has a historical momentum, a life of its own, as it were. It appears to be impervious to policy innovations. The most difficult problem in S&T policy formulation seems to be the cultivation of the receptivity of the political leaders and the science community itself to new policy directions.

The economic history of the Philippines is one of chronic crisis and increasing poverty for an increasing number of its citizens. This is reflected in the continous devaluation of the Philippine peso from 2 pesos to the US dollar in 1946 to 20 pesos to the dollar in 1986. Poverty has also inexorably increased. Forty years ago, about 40 per cent of the population was below the poverty line. Today, the figure could be as high as 85 per cent.[7] Although the economy showed impressive gains in the three decades after the Second World War,[8] the lasting foundations for sustainable growth in terms of indigenous capability to support modern productive processes were not put into place. In spite of the respectable increase in Net Domestic Product in the last few decades, Philippine economic development was accompanied by high rates of unemployment, wide disparities in income distribution, and regional concentration of productive facilities in the Metro Manila area.

During the period 1949–1969, the annual average growth of manufacturing was an impressive 8.5 per cent. However, this was mostly illusory industrialization, because there was negligible enhancement of local technological capability. Furthermore, the policy merely favoured the manufacture of import-substituting consumer goods and discriminated against the manufacture of capital goods and exports.

The lack of concern for the technological aspects of industrialization is also reflected by the expenditures of private industry on R&D. In a survey of the 50 largest industrial firms in 1956, it was found that only 350,000 pesos were spent for R&D.[9]

The early exuberant growth of manufacturing in the Philippines in the 1950s was not sustainable because it was not self-reliant growth. It was not based on local technological capability, and it did not rely on local innovation and international competitiveness for growth. It was simple import-substitution with most of the capital goods and technology imported.

Dependence on foreign technology is apparent in the predominance of foreign brands in the Philippine consumer market. To a certain extent, this was the result of the early industrialization efforts based on an undiscriminating import-substitution policy. The case of the cigarette industry provides an interesting insight. Before the 1950s, there were many local brands of cigarettes. When the American brands were introduced in the 1960s, local brands were pushed into oblivion. What eventually survived were those companies with strong links to American companies.

Because of the failure of the policies of the 1960s to stimulate the industrial sector, a package of attractive incentives was put together in the form of the Industrial Incentives Act of 1967. The Act was aimed at stimulating investment in industrial enterprises, and consisted primarily of fiscal inducements. In subsequent legislation (P.D. 92), fiscal incentives were also provided to promote labour intensiveness and backward integration.

Yet these incentives did not include a provision on the transfer of technology and the assurance of a learning process for local technologists. Although a Technology Transfer Board was created, the main motivation was to safeguard the interests of local investors and not to ensure a real technology transfer.

In spite of the Industrial Incentives Act and its obvious attractions, industrial growth was very modest in the 1970s (table 3). Only a few took advantage of the government incentives. In 1973, there were only 131 firms registered with the government.[10] It was clear that entrepreneurs had no real interest in new, pioneering ventures – an attitude which persists up to the present.

The uneven growth of agriculture since the 1950s (table 2) and the lack of a outstanding performance indicates that the relatively heavy emphasis on agricultural R&D was not significant.

The general picture of S&T manpower in the Philippines is shown in table 4. In this table, the 1965 data, obtained from the Survey of Scientific and Technological Manpower conducted by the NSDB in 1965, serve as the baseline figures. Using these figures, the future supply of S&T manpower was calculated from the data on graduates of the country's educational system.

The following observations could be made from the manpower figures:

1. The proportions of scientists and engineers increased only slightly, from 26.5 per 10,000 in the 1960s to 27.0 per 10,000 in the 1970s. This is expected to remain more or less constant up to the 1990s. The growth rates are quite low. The numbers appear to be rather high when compared to more developed countries. However, a

Table 3. Average annual growth rates of domestic product (1972 prices) by industrial origin, 1949–1982 (percentages)

	1949–53	1953–57	1957–61	1961–65	1965–69	1969–73	1973–77	1978–82
Agriculture	7.7	4.3	4.2	4.6	4.0	3.4	5.4	4.1
Industrial sector	8.8	8.1	3.7	5.8	5.5	7.3	8.1	4.9
Mining	23.5	7.7	1.0	2.7	14.6	11.4	4.3	3.2
Manufacturing	14.1	11.1	5.7	4.8	6.6	7.5	5.0	3.8
Construction	0.3	2.6	-1.6	10.8	-0.6	5.2	21.8	8.7
Utilities	3.6	5.7	2.5	3.0	5.3	7.9	11.2	9.0
Service sector	9.4	0.6	4.6	4.6	4.7	4.6	5.2	4.7
Net domestic product	8.6	6.2	4.2	4.8	4.6	4.9	6.1	4.3

Source: National Accounts Staff, National Economic and Development Authority.

Table 4. S&T manpower in the Philippines

Year	Population (in units)	Total stock (in units)	Scientists and engineers Total number (in units)	Working in R&D Natural sciences	Breakdown by field of education training (units) Engineering and technology	Agricultural sciences	Medical sciences	Social sciences
1965	31,886,081	81,600	25,600	1,838	11,900	2,187	5,575	4,100
1970	36,684,981	100,385	31,493	2,261	14,638	2,690	6,858	5,044
1975	42,070,660	115,482	36,230	2,601	16,840	3,094	7,890	5,802
1979	47,719,000	128,897	40,433	2,903	18,793	3,453	8,805	6,476
1985	53,108,000	157,000	–	–	–	–	–	–
1990	59,846,000	177,026	–	–	–	–	–	–

Source: UNESCO.

small fraction of the total S&T manpower has graduate training and fewer still are engaged in R&D.

2. In the natural sciences, there were marked increases. For the 1980s, increases in all fields are predicted except in agriculture. However, growth rates will decrease.
3. The trend shows that the number of scientists and engineers in R&D will increase, but at a decreasing rate.

There were 1,157 colleges and universities in the Philippines in 1984. About 73 per cent of these were private schools. Of the graduates of these schools, classified by field, only 22 per cent studied engineering and the sciences.[11]

The state of science education in the country is perhaps reflected by the figures in physics, chemistry, and mathematics. In a survey conducted by the Kilusan ng mga Siyentipiko sa Pilipinas, the following facts were reported:

1. In the 1970s, there were 250 chemists. Only 15 per cent of them had graduate degrees.
2. In physics, there were 21 Ph.D.s and 15 M.Sc. degree-holders. Only 10 were actively engaged in research. More than 95 per cent of college physics teachers did not have a B.Sc. physics degree.
3. In mathematics, fewer than 1 per cent of the teachers had Ph.D. degrees.
4. There were only 10 institutions which were doing research. The University of the Philippines had 75 per cent of the total research projects.

Tables 5 and 6 reflect the allocation of financial resources for R&D. Three general sources are: the government, the private sector (private industry and foundations), and foreign sources.

Table 5. R&D expenditures: Breakdown by source and sector of performance, 1979 (in thousands of pesos)

Sector of performance	National		Foreign	Total
	Government funds	Other funds		
Productive	820	75,815	876	77,511
Higher education	84,250	49,347	–	133,597
General service	166,914	53,871	14,148	234,933
Total	216,372	143,579	7,830	367,781

Source: UNESCO, *Science and Technology in Countries of Asia and the Pacific: Policies, Organization and Resources*, Paris: UNESCO, 1985.

Table 6. R&D expenditure: trends

Year	Population (millions)	GNP (in millions of pesos)	Total R&D expenditure (in thousands of pesos)	R&D expenditure as percentage of GNP	Per capita R&D expenditure
1965	31.886	23,382	41,198	0.18	1.29
1970	36.685	41,751	65,056	0.16	1.77
1975	42.071	114,265	239,233	0.21	5.69
1979	47.719	220,935	446,041	0.20	9.35
1985	53.108	475,147	985,455	0.21	18.56
1990	59.846	979,089	2,046,748	0.21	34.20

Source: UNESCO, *Science and Technology in Countries of Asia and the Pacific: Policies, Organization and Resources*, Paris: UNESCO, 1985.

Table 7. Per capita government budget for education

Schools	Cost per student (1982 pesos)
Public elementary schools	392
National secondary schools	1,037
Locally funded high schools	123
Government tertiary schools	5,636

Source: MECS and NSTA, Science Education Development Plan, Vol. 1, November 1985.

It is worth noting that in 1979 about 69 per cent of the total R&D funds were provided by the government. The private sector contributed about 39 per cent, with foreign sources contributing about 2 per cent. The total amount spent for R&D was 367 million pesos (US$50 million). As seen in table 6, the expenditure on R&D as a fraction of the GNP is not expected to change significantly up to 1990.

Looking at trends in R&D expenditure in the decade ending in 1975, there was no significant change in terms of US dollar values. In 1983–1985, there was a significant decrease in the US dollar values of R&D expenditures because of inflation.

In regard to expenditure in education at the tertiary level, the 313 government colleges and universities were allocated 717 million pesos in 1983–1984. The country's investment in education is shown in table 7.

The institutions engaged in S&T activities are shown in figure 4. Three categories are used: R&D agencies, S&T education and training, and S&T services and delivery.

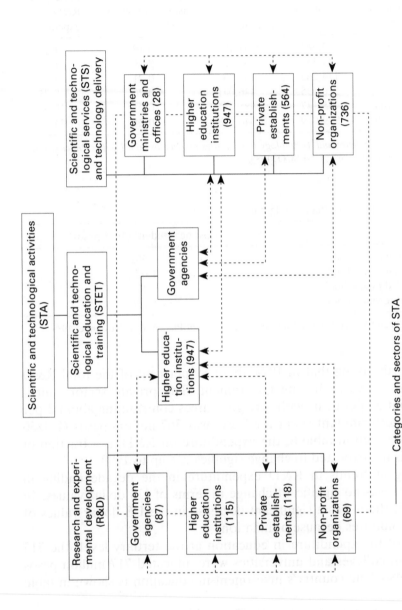

Fig. 4. Institutional network (Source: UNESCO)

——— Categories and sectors of STA

------- Linkages between sectors

Numbers in parenthesis refer to the number of institutions performing STA

The mechanisms and interrelationships in this network are as follows:

1. The R&D institutions are linked through the NSTA and its policy councils. R&D in the universities and colleges is coordinated through the grants-in-aid programme of the NSTA.
2. The private and government educational institutions produce the manpower for the R&D and S&T services and delivery systems.
3. In S&T services, those provided by the private sector are also utilized by the government. On the other hand, the information services of the universities are used by both the government and the private sector.
4. In the delivery of S&T services, government agencies that have commercially viable technologies are assisted in contacting private industry for possible ventures.

The principal agency for S&T development is the National Science and Technology Authority (NSTA), whose organizational chart is shown in figure 5. While the NSTA attempts to centralize S&T activities, almost the entire government bureaucracy is involved in one way or another.

The R&D of various research institutes is intended to be coordinated by the various councils of the NSTA. Except in the case of agriculture and natural resources, this has not been very successful.

In 1978, a "consortium concept" was introduced by the NSTA. The general idea was to pool the resources of the various S&T units, particularly the universities. In 1983, another concept called "science communities" was included in the growing lexicon of S&T in the Philippines. Four science communities were established: the Bicutan, Diliman, Ermita, and Los Banos. These communities are groupings of research and academic institutions that are in physical proximity to each other. They are expected to promote an environment of productive and creative interaction and cooperation among the members. Housing and other social amenities are supposed to be provided. Sharing of facilities and resources is encouraged.

In the Philippines, foundations registered with the NSTA enjoy some privileges, such as tax exemptions, provided they do scientific and technological R&D. In 1979, there were 69 foundations registered. The top three in terms of R&D expenditure are the Population Centre Foundation, the Philippine Business for Social Progress, and the Filipinas Foundation. None of these is engaged in R&D in the hard sciences. In private industry, there were 118 private firms engaged in R&D in 1980. Most of the R&D was in the areas of textiles, paper products, food, beverages, tobacco, and chemicals.

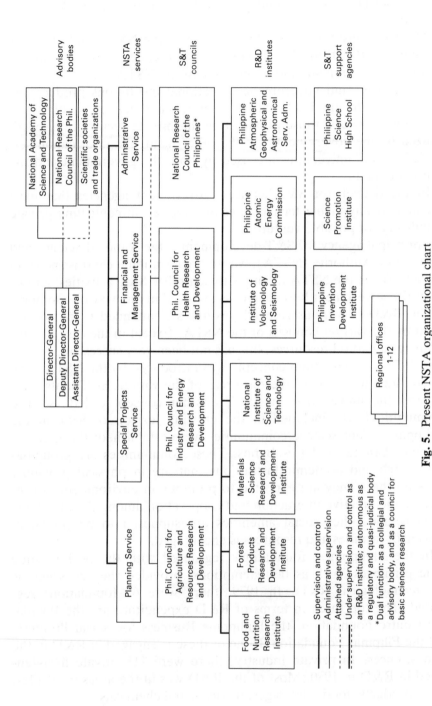

Fig. 5. Present NSTA organizational chart

Case-studies

The primary objective of the case-studies was to test the validity of the assertions and hypotheses arising out of the historical analysis. They were also expected to provide new insights into the dynamics of S&T and social development in the real world. Whenever possible, the case-studies attempted an empirical determination of micro-level indicators. At the very least, the indicators would provide a frame of reference for the interviews of principal actors and for the ensuing analysis.

The need for a traditional qualitative approach which involves intuitive and normative processes to achieve insights becomes more convincing when used in parallel with the indicator-empirical approach. The two approaches are complementary and mutually reinforcing. Since the problem involves numerous variables, an unconstrained freedom of imaginative exploration could yield valuable insights and discoveries. However, the standard norms of the scientific process must not be unduly compromised if we are to claim any degree of scientific validity. This is the rationale for using two parallel and simultaneous approaches in the case-studies.

The analytical framework for the case-study is shown in figure 7. The case-studies were directed at the following sectors:
1. The copper industry.
2. The alternative energy sector.
3. The coconut industry.
4. The semiconductor industry.

The copper industry is one of the oldest industries in the Philippines. It was hoped that it could provide significant insights into the evolution of technological capacity. The recent establishment of a local smelter and refinery would necessarily involve new technological inputs and backward linkages to the copper-wire industry. At the same time, the validity of the copper R&D establishment is being challenged by competition from aluminium and fibre optics. The response of the industry to its present depressed state and to the threat of technological obsolescence could reveal the salient structural features of its technological capacity.

The energy crisis of 1973 found the Philippines heavily dependent on imported oil. Ninety-five per cent of its energy requirement was supplied by imported petroleum. A crash programme to attain partial self-sufficiency was undertaken with strong government support. The non-conventional alternative energy sector, principally the geothermal and biogas systems, was considered a priority area.

Because of a strong political commitment to geothermal energy, the sector is an interesting subject for a case-study. And, the role of political support could be tested empirically.

The biogas subsector is also considered a good case-study because it is one of the very few examples of a successful response to the energy crisis by a private company through R&D. The biggest biogas facility in the Philippines (Maya Farms) is the product of more than a decade of R&D.

Like the copper industry, the coconut industry has a long history in the Philippines. It started in 1768 when a Spanish decree ordered the planting of coconuts. The export of copra began in 1895, and the first commercial oil mill was constructed in 1906. However, what makes the industry interesting for a case-study is the long history of R&D, which is mostly government-supported.

The semiconductor industry is claimed to be the Philippines' venture into high technology. The industry presents an interesting mix of subsidiaries, joint ventures, and fully Filipino-owned firms. A case-study of this sector could reveal the features of various modes of technology transfer and the role of equity.

In terms of the S&T taxonomical matrix, the choices for the case-studies are shown in table 8. In this sense, they are quite representative of the total technology-industry picture in the Philippines.

The essential input to the case-study consists of the following background information:
1. List of companies involved in the sector.
2. Company profiles.
3. Profiles of engineers.

Table 8. S&T taxonomical matrix for the case-studies

Type of technologies	First-wave technologies	Second-wave technologies	Third-wave technologies
Materials technologies		Copper industry	
Equipment technologies			
Energy technologies		Geothermal energy	
Information technologies			Semiconductor industry
Life technologies	Coconut industry		

4. Academic/government institutions involved in R&D in the sector.
5. List of training and educational institutions relevant to the sector.
6. Statistics on manpower with relevant expertise involved in the sector.
7. Compilation of literature on the economics of the sector.
8. Compilation of legislation and government policies and regulations on the sector.
9. Listing of relevant regulatory agencies.
10. Listing of the principal actors in the sector.
11. Compilation of historical literature on the sector.
12. Compilation of technology transfer arrangements in the sector.
13. Compilation of the local patents on the sector.
14. Listing of components of the techno-system.

Most information of significance is not in documented form. This is especially true of technology transfer arrangements and pricing. Interviews and commissioned background papers are primarily intended to overcome this constraint. Symposia and workshops involving the leading actors could yield further information and insights. Thus, documented information from bibliographic research supplemented by commissioned papers and the results of workshops are two major sources of inputs, illustrated in figure 6. These two sources are used to evaluate the self-reliance indicators, which are defined below. In addition, these are also utilized in determining the role of exogenous forces such as political, social, cultural, economic, and technology transfer factors.

For each case-study, two types of analysis are used. The cross-impact analysis is simply a semi-quantitative evaluation of the influence of the various exogenous factors on the set of self-reliance indicators. This will serve to identify the factors with most influence on particular dimensions of self-reliance.

The normative analysis is essentially qualitative, that is, complementary and supplementary to the cross-impact analysis. It involves the assessment of the explicit goals of the techno-system *vis-à-vis* the self-reliance indicators. In addition, sectoral goals are analysed in the context of national goals and needs.

In general, the research methodology presented here is a logically structured approach. It is a synergistic mixture of quantitative and qualitative elements.

The first step in the process of getting an operational fix on the concept of technological self-reliance at the micro level is to enumerate all the relevant characteristics of the techno-system. These characteristics

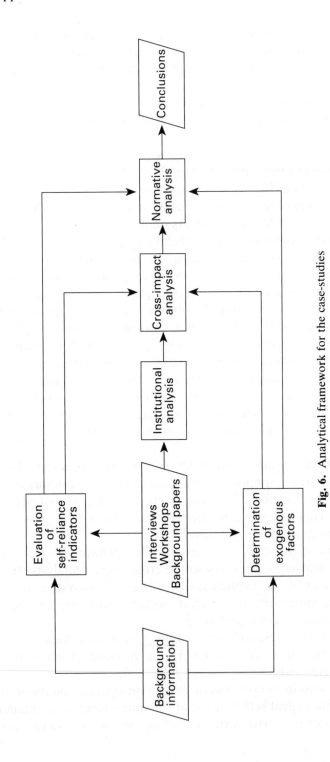

Fig. 6. Analytical framework for the case-studies

216

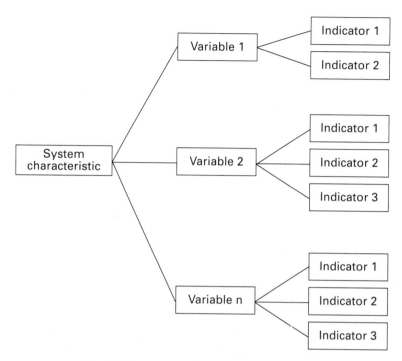

Fig. 7. The process of quantifying self-reliance

are subsequently expressed in terms of variables that relate to self-reliance, and each variable is further broken down into measurable indicators. The process is depicted in figure 7. The result of this process is shown in Appendix 1.

Each indicator will, of course, be chosen to be theoretically measurable. One way of doing this is to choose indicators that could be expressed as an ordinal set of rank order categories. The ordinal set must be constructed in a standard undirectional classificatory principle. Here it is conceptualized as a continuum from absence to presence, and from low to high, as illustrated below.

Indicator values	*Empirical reference*
1	Absent, low
2	Partly, medium
3	Present, high

Each indicator should have a special and suitable formulation of empirical reference. The result of the process is shown in Appendix 1, where the system characteristics, variables, and indicators are listed.

In the actual studies, the relevant indicators are evaluated through the analysis of background information and interviews. The evaluation of the indicators reveals the most significant dimensions of self-reliance. It must be emphasized that the operational definition presented here is primarily intended to fix the meaning of the multi-dimensional concept of S&T self-reliance. The numerical values of the indicators are relatively unimportant compared to the semantic clarification of the concept. Moreover, the concept of the techno-system and the self-reliance indicators provide a common framework for the different case-studies. The ultimate prize was that the case-studies could be undertaken in a uniform systematic manner. Consequently, the results have a high degree of comparability.

Case-study results

Copper industry

We considered the techno-system for copper wires and cables, copper cathodes, and copper concentrates, as shown in figure 8. For simplicity's sake we will refer to this techno-system as the copper industry. The evaluation of self-reliance indicators for the copper industry was undertaken through the analysis of existing literature and background papers and the interviews of the principal actors of the sector. The results are shown in table 9.

The important dimensions of S&T self-reliance are obvious. The general assessment is that in terms of technological capability, the system is still in the operative stage, with some indications of nascent adaptive capability. This is confirmed by the value of indicator no. 4.21 in table 9.

The self-reliance of the copper industry is weak in the following dimensions, where the average score is 1.5 or less:
- Quality of technological innovation by Filipinos.
- Support of local R&D by the industry.
- Utilization of R&D results.
- Local supply of hardware.
- Number of innovations in the industry.
- Control of financing.
- Foreign nationals in management.

Both the quality and the quantity of local innovations in the industry are unsatisfactory. Of the 10 important patents registered for the industry, only one is of Filipino origin.

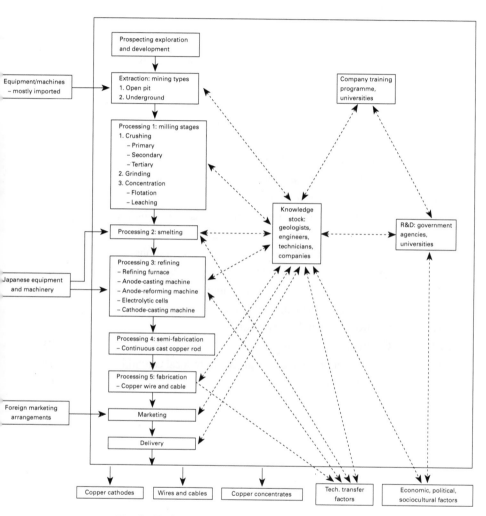

Fig. 8. Techno-system for the copper industry

219

Table 9. Indicators of self-reliance in science and technology: copper industry

Indicator number	Indicator	Average values
3.22	Change in the number of Filipinos with managerial know-how	3
6.23	Local maintenance of hardware	2.6
6.12	Adequacy of number of graduates	2.6
1.11	Existence of technical training programme in corporate plans	2.6
7.11	Existence of the various components of the techno-system for product X	2.5
3.21	Change in number of Filipinos with the technical know-how in relevant technologies	2.4
2.21	Control of managerial inputs	3.25
3.13	Use of local material inputs to the various processes	2.2
1.31	Role of nationals in policy formulation	2.2
5.21	Utilization of locally trained technicians and engineers	2.2
4.12	Existence of historical industry statistics	2.2
2.12	Equity of participation of nationals in corporations	2
4.11	Existence of technical industry library locally	1.8
1.21	Existence of plans for local autonomy	1.8
6.11	Relevance of curricula to the industry	1.8
2.22	Control of technological inputs	1.75
2.23	Control of material inputs	1.75
3.14	Adaptation of some of the processes to local conditions	1.75
6.13	Quality of the graduates	1.6
2.24	Control of financing	1.5
1.12	Existence of R&D programme in corporate plan	1.5
2.11	Nationality of management	1.5
1.22	Plans for vertical integration	1.33
3.11	Number of innovations in the industry	1.25
6.22	Local supply of hardware	1.2
4.21	Technological capacity	1
5.11	Support of local R&D by industry	1
5.12	Utilization of R&D results	1
3.12	Quality of technical innovations by Filipinos	1

In R&D, the sector is not doing anything important. The industry's only significant link to government R&D is in the area of geological exploration. The interviews with the technological leaders of the industry revealed a static perspective on process or product improvements. The prevailing general attitude is that R&D cannot help the industry except in trivial ways.

In spite of the constitutional provision that natural resources-based industries should be at least 60 per cent owned by Filipinos, alien control of the copper industry is still significant. This is indicated by the following facts:[12]

– The key officers of the biggest firms are mostly foreigners.
– The biggest sales contracts are with a very few foreign firms.
– Of the combined resources of the 10 biggest mining companies – about P22 billion – only P9.8 billion is owned by stockholders; the rest is owned by creditors.
– Only about P5.9 billion of the mining resources, representing 27 per cent of total assets, are owned by Filipinos; 73 per cent are owned by foreign stockholders and leaders.

The strongest aspects of self-reliance (indicator value greater than 2.5) are the following:

– Local maintenance of hardware.
– Technical training programmes in companies.
– Adequate supply of graduates.
– Vertical linkages of the system.

Although almost all the capital equipment is imported, local technicians can maintain these properly. This further reinforces the observation made previously that the industry is still in the operative state and has not reached the replicative stage. This is consistent with the finding that there is an adequate supply of engineers and technicians but they are of low quality.

In summary, then, the copper industry is technologically dependent. The historical, political, and social factors conspire to create this condition of perpetual dependency. The recent closures of many firms in the industry are perhaps symptomatic not only of dependency but of a latent potential for self-destruction.

Geothermal energy

The techno-system for the geothermal energy sector is shown in figure 9. There are four processing stages, starting from the geophysical and geochemical survey and leading to the transformation process in which

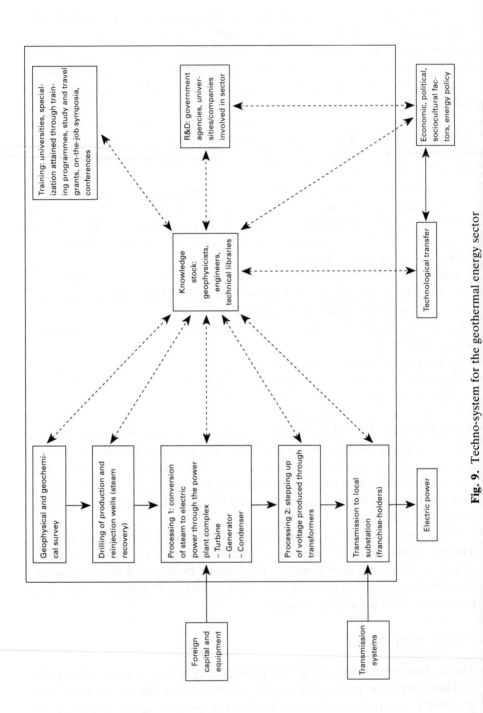

Fig. 9. Techno-system for the geothermal energy sector

Table 10. Indicators of self-reliance in science and technology: alternative energy sector (geothermal and biogas)

Indicator number	Indicator	Average values	
		Geothermal	Biogas
1.11	Existence of technical programme in corporate plans	3	3
3.21	Change in number of Filipinos with technical know-how in relevant technologies	3	3
3.22	Change in the number of Filipinos with managerial know-how	3	3
4.11	Existence of technical industry library locally	3	3
4.12	Existence of historical industry statistics	3	3
4.3	Number of scientists, engineers, and technicians in relevant fields	3	3
5.21	Utilization of locally trained technicians and engineers	3	3
6.11	Relevance of curricula to industry	3	3
6.12	Adequacy of number of graduates	3	3
7.11	Existence of the various components of the techno-system for product	3	3
7.12	Interdependence/linkage of subsystems	3	3
1.21	Existence of plans for local autonomy	2	3
1.22	Plans for vertical integration	2	3
1.31	Role of nationals in policy formulation	2	3
2.12	Equity of participation of nationals in corporations	2	3

the voltage output is stepped up before it is transmitted to the local substation.

Interviews with some of the principal actors in the sector and a thorough analysis of literature and background papers written about geothermal energy operations in the Philippines facilitated the evaluation of the self-reliance indicators. The results can be seen in table 10. From these, it can be generalized that the Filipinos are more or less capable as regards the main technologies involved in a full-scale geothermal energy production. Skills in geothermal exploration and drilling management have been acquired by a significant number of Filipinos. These skills range from the evolvement of design layouts for steam collection and effluent disposal systems to the supervision of their construction and the eventual maintenance of the systems. In terms of technological capability, therefore, it can be surmised that the system has clear indications of adaptive capability.

It is worth noting that technology transfer was assured here through the training of local professionals under "hands-on" conditions. This was provided for in the contract with Union Oil. This training programme was complemented by bilateral training assistance programmes with New Zealand, Italy, Japan, and the United States. By 1985, there were more than 280 technical personnel trained in other countries.[13]

As can be seen in table 10, self-reliance is low or absent in the following dimensions, where the score is 1:
– Existence of R&D programme in corporate plans.
– Nationality of management.
– Number of innovations in the industry.
– Quality of technological innovations by Filipinos.
– Utilization of R&D results.
– Local supply of hardware.
The quantity of local innovations in the industry is unsatisfactory. There are only two important patents registered, neither of which is Filipino in origin.

Self-reliance is very strong in the following areas, where the indicator value is 3:
– Existence of technical training programme in corporate plans.
– Number of Filipinos with managerial know-how.
– Existence of technical industry library locally.
– Existence of historical industry statistics.
– Utilization of locally trained technicians and engineers.
– Relevance of curricula to industry.
– Adequacy of number of graduates.
– Local maintenance of hardware.
– Existence of the various components of the techno-system.
– Interdependence/linkages of the subsystems.
From all indicators, the geothermal sector, being a very important energy source, has been given more than adequate attention by the government. However, the government has yet to make it more viable for local geothermal companies to operate in the Philippines. Laws on tax exemptions on capital goods and other inputs, plus presidential proclamations setting aside geothermal areas as national reservations, are indicators of the heavy government support for the sector.

The geothermal energy sector is still technologically dependent on imported capital equipment. However, in spite of this, there are clear indications that it has achieved a semblance of self-reliance in its man-power requirements and in the adaptation of the technology to local conditions.

The existence of relevant indigenous R&D and strong government support for the industry's development account for the higher degree of technological capability in this field. It is unfortunate, however, that this thrust is not being pursued with a sustained commitment to reach the level of technological mastery.

Coconut industry

The major products derived from coconut trees are copra, coconut oil, copra meal/cake, and dessicated coconut.

Figure 10 shows the six processing stages, starting from the planting and cultivation of coconut trees to the processing of coco-chemicals. The knowledge stock of the industry is with engineers, chemists, botanists, agriculturalists, technicians, and managers. Training is carried out mainly in the universities and through company-sponsored training programmes. R&D is undertaken by government agencies and universities. The government agencies are the Philippine Coconut Authority and the Bureau of Plant Industry, and the universities UP Los Banos, Palawan National Agricultural College, the University of Eastern Philippines, Visayan State Agricultural College, and the University of San Carlos.

The results of the evaluation of self-reliance indicators are shown in table 11.

With respect to coconut milling and refining, the system is in the replicative stage of technological capacity. However, at the coco-chemical processing stage, it is still at the operative level. The average value of indicator no. 4.21 is 1.5, indicating that the system is weak in terms of technological capacity.

Other indicators show that the sector is weak in the following aspects of self-reliance:
- Existence of technical training programmes in corporate plans.
- Quality of technical innovations by Filipinos.
- Utilization of R&D results.
- Level of R&D effort.
- Support of local R&D.

There are no explicit technical training programmes in the industry's corporate plans. The industry, then, does not put emphasis on learning in its formulation of goals.

Table 11. Indicators of self-reliance in science and technology: coconut oil industry

Indicator number	Indicator	Average values
4.12	Existence of historical industry statistics	3
4.3	Number of scientists, engineers, and technicians in relevant fields	3
5.21	Utilization of locally trained technicians and engineers	3
6.12	Adequacy of number of graduates	3
6.22	Local maintenance of hardware	3
1.31	Role of nationals in policy formulation	2.5
2.11	Nationality of management	2.5
2.12	Equity of participation of nationals in corporations	2.5
2.21	Control of managerial inputs	2.5
3.13	Use of local material inputs to the various processes	2.5
4.11	Existence of technical industry library locally	2.5
7.11	Existence of the various components of the techno-system for product	2.5
1.12	Existence of R&D programme in corporate plans	2
1.21	Existence of plans for local autonomy	2
1.22	Plans for vertical integration	2
2.22	Control of technological inputs	2
2.23	Control of material inputs	2
2.24	Control of financing	2

The technical innovations by Filipinos are of little value to the industry. The industry does not support local R&D and utilizes hardly any local R&D results.

The strongly self-reliant aspects of the industry are the following:
- Role of nationals in policy formulation.
- Nationality of management.
- Equity participation of nationals in corporation.
- Control of managerial inputs.
- Use of local material inputs to various processes.
- Existence of technical industry library locally.
- Existence of historical industry statistics.
- Number of scientists, engineers, and technicians in relevant fields.
- Utilization of locally trained technicians and engineers.
- Adequacy of the number of graduates.

– Local maintenance of hardware.
– Existence of the various components of the techno-system for coconut.

Control of the techno-system rated high self-reliance scores. Philippine nationals have increased their participation in the industry's control.

In the past, while actual crop cultivation was in the hands of Filipinos, the intermediate processing of coconut products for export and local distribution was dominated by foreign capitalists. In recent years, a new breed of local capitalists, composed of coconut landlords, Filipino capitalists, and government bureaucrats, has emerged and taken over the processing and export of coconut oil. Foreign capital, however, has remained in the manufacture of coco-chemicals.

In terms of material inputs, the industry has increasingly made use of local materials in the various processes. Local maintenance of hardware is ably done by local technicians. In terms of local manpower there is an adequate supply of engineers and technicians for the industry.

Foreign technological and material inputs continue to flow into the various sectors of the industry. Although the industry has made moves to disengage itself from foreign control, it continues to follow a colonial pattern of trade. Its present orientation still maintains the colonial and agrarian character of the Philippine economy. The refining and manufacturing of coconut-based consumer products are still in the hands of TNCs.

Semiconductors

Over a span of 10 years, the semiconductor industry experienced tremendous growth. There were 17 companies in 1978, and these grew to 33 foreign and local firms in 1981. In a matter of nine years since the first shipment of electronic components in 1973, worth US$10 million, semiconductors have become the Philippine's top non-traditional export. Table 12 shows the performance of semiconductors in relation to other prime exports. The techno-system for the semiconductor industry is shown in figure 11.

The Philippine semiconductor plants only carry out assembly, and are classified into: (a) captive producers or foreign-owned subsidiaries who turn out products for the exclusive use of the mother company; and (b) independent contractors who cater to the requirements of various customers.

Raw materials are consigned to subsidiaries by the mother company

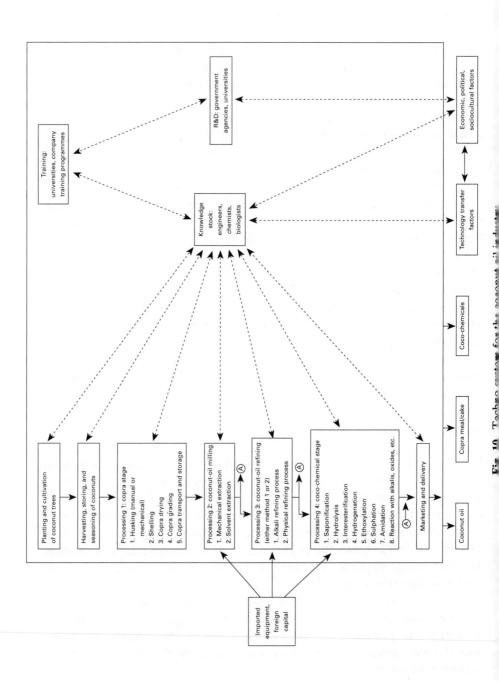

Fig. 10. Techno system for the coconut oil industry

Table 12. Top commodity exports, 1980–1983 ($US millions)

Rank	1980		1981		1982		1983	
	Commodity	Value	Commodity	Value	Commodity	Value	Commodity	Value
1	Sugar	624	Sugar	566	Sugar	416	Coconut oil	516
1	Coconut oil	567	Coconut oil	533	Coconut oil	401	Garments	409
3	Copper concentrates	545	Garments	458	Garments	397	Sugar	299
4	Garments	362	Copper concentrates	429	Semiconductors	329	Semiconductors	256
5	Gold	239	Semiconductors	271	Copper concentrates	312	Copper concentrates	249
6	Semiconductors	179	Gold	215	Gold	169	Gold	154

Sources: NCSO Foreign Trade Statistics; External Trade Statistics Group, DER International; Special Study, Technical Staff, Foreign Exchange Committee.

a. Garments are net of import value of material inputs for exports on consignment basis, while semiconductor devices are net of import value of material inputs for consigned exports and payments for loans and advances from parent companies.

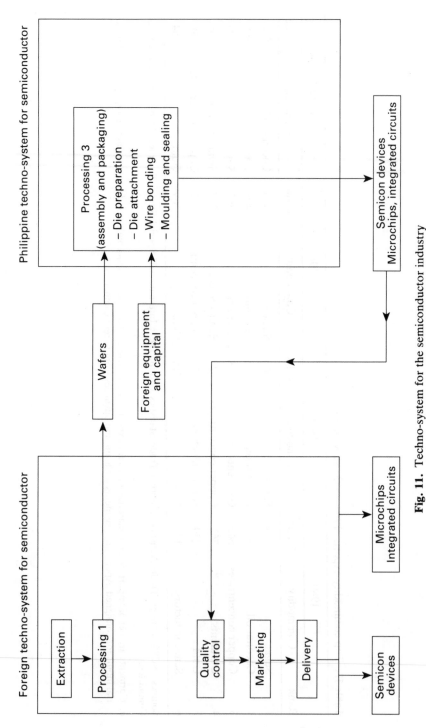

Fig. 11. Techno-system for the semiconductor industry

or to subcontractors by the foreign clients. These imported raw materials include silicon dice, wafers, metal can packages, aluminium wire, gold wire, epoxy caps and bases, chemicals, etc. In packaging assembly, there are four stages which are done in the Philippines; a few companies have testing facilities. All marketing activities are done by the mother company/customer.

For foreign-owned subsidiaries, R&D activities are carried out by the mother company. Minimal innovations (e.g. simplification of word processes) are made in the Philippines. An evaluation of self-reliance indicators, based on existing literature, plant visits, and interviews with key officers in the industry, is shown in table 13.

In foreign-owned subsidiaries, the chief executive officers are mostly expatriates, except for Timex, which is headed by a Filipino. Filipino engineers occupy managerial and supervisory positions, as the country has an adequate supply of skilled and highly trainable technical personnel. Technical training programmes are present in most companies as part of the regular manpower orientation and training programme.

R&D and linkages with relevant institutions are non-existent. Self-reliance is weak in the following aspects;
– Existence of R&D programmes.
– Control of technological inputs.
– Control of material inputs.
– Control of financing.
– Innovations in the industry.
– Existence of technical industry library.
– Local supply of hardware.
The Philippines does not in fact have a "semiconductor industry." The labour-intensive phase of production is carried out here primarily because of inexpensive labour, whereas the more capital- and technology-intensive stages take place in the mother companies or clients' facilities. There is a need to develop the industry's backward and forward linkages to establish technological independence.

The industry could break out of this situation by: (1) establishing local R&D; (2) training the Philippines' unlimited labour force for high technologies; (3) providing investment and financial assistance from local sources; (4) developing the support of allied industries; (5) developing new products; (6) expanding and seeking new markets; and (7) promoting regional and global cooperation through exchanges of technologies.

Self-reliance may be achieved by moving in the following directions:
1. Establishing a semiconductor R&D centre.

Table 13. Indicators of self-reliance in science and technology: semiconductor industry

Indicator number	Indicator	Average values
3.21	Change in number of Filipinos with technical know-how in relevant technologies	3.0
1.11	Existence of technical programmes in corporate plans	2.5
3.22	Change in the number of Filipinos with managerial know-how	2.5
4.30	Number of experts	2.5
5.21	Utilization of locally trained technicians and engineers	2.5
6.12	Adequacy of number of graduates	2.0
6.13	Quality of the graduates	2.0
6.23	Local maintenance of hardware	2.0
2.21	Control of managerial inputs	2.0
3.13	Use of local material inputs to the various processes	2.0
3.14	Adaptations of some of the processes to local conditions	2.0
1.31	Role of nationals in policy formulation	1.5
2.11	Nationality of management	1.5
4.12	Existence of historical industry statistics	1.5
6.11	Relevance of curricula to industry	1.5
7.11	Existence of the various components of the techno-system for product X	1.5
1.12	Existence of R&D programme in corporate plans	1.0
1.21	Existence of plans for local autonomy	1.0
1.22	Plans for vertical integration	1.0
2.12	Equity of participation of nationals in corporations	1.0
2.22	Control of technological inputs	1.0
2.23	Control of material inputs	1.0
2.24	Control of financing	1.0
3.11	Number of innovations in the industry	1.0
3.12	Quality of technical innovations by Filipinos	1.0
4.11	Existence of technical industry library locally	1.0
4.21	Technological capacity	1.0
5.11	Support of local R&D by industry	1.0
5.12	Utilization of R&D results	1.0
5.30	Level of R&D effort	1.0
6.22	Local supply of hardware	1.0
7.12	Interdependence/linkage of subsystems	1.0

2. Organizing a national association of semiconductor producers in the Philippines.
3. Creating a government agency that would protect local interests in the industry.
4. Setting up a Philippine wafer fabrication facility.
5. Developing a semiconductor industry complex that would spearhead the transformation of the industry from more offshore assembly houses to a total manufacturing base for semiconductor production.

Technological dependence: Nature and consequences

Technological development is essentially a historical process that depends on a society's initial state of S&T knowledge and the significant influences that modify, enlarge, and stimulate that existing base. When the Philippines first made contact with the more technologically developed Western cultures, its technological capabilities and capacity were primitive. Because of the very unequal development of the two cultures, Western domination was the inevitable result. The contact did not result in any significant technological learning for the Philippines over the centuries of Spanish and American rule. After more than 400 years of colonization, the Philippines still relies on other countries, principally the US and Japan, for most of its technology. This situation is a sure symptom of technological dependence.

Like most multidimensional concepts, technological dependence is difficult to define rigorously. A clarification is attempted here.

Technological dependence could be considered the opposite of self-reliance. One transparent indicator of its existence is a situation in which the major source of a country's technology is abroad.[14] When a country imports from a single country, it has a very high degree of technological dependence. From these considerations alone, one is easily convinced of the state of technological dependence of the Philippines.

There are other useful macro-indicators of technological dependence, which are shown in table 14. The consistently low figures for the number of scientists and engineers, R&D expenditures and patent grants for the Philippines, compared to the Republic of Korea and Japan, reflect its undeniable state of underdevelopment and technological dependence. There is no desire to belabour this point. However, the various indicators show the particular areas of relative weakness. These could be useful in the formulation of policies and strategies.

Table 14. Technological capacity: selected indicators, 1982

	Japan	Republic of Korea	Philippines
Total R&D personnel	648,977	46,390	17,992
Scientists and engineers engaged in R&D per 10,000 population	40.44	7.23	1.53
Technicians engaged in R&D per 10,000 population	7.68	2.97	0.69
Expenditure on R&D as percentage of GNP	2.5	0.9	0.2
Application for patents filed by residents[a]	227,708	1,599	63
Grants of patents to residents[a]	45,578	245	52

Sources: *UNESCO Annual Statistical Yearbook*, 1984; Techno-economic Evaluation Division, Planning Service, NSTA; ADB, *Key Indicators of Developing Member Countries of ADB*, 1985; Bank of Japan, *Economic Statistics Manual*, 1984; *Philippine Yearbook*, 1983.

a. 1983.

The case-studies provided a very detailed picture of the gross features and nuances of self-reliance in particular industries. Table 14 summarizes the strong and weak points with respect to the self-reliance micro-indicators arising from the results of the case-studies.

It is interesting to observe that in the older industries like copper and coconut, the weak points in self-reliance are those relating to inadequate and low quality of R&D. The prognosis here is that there is very little desire and ability to improve the industries technologically. In spite of their long histories, technological capability has remained low. In general, the capabilities are still in the operative stage. These observations are consistent with the very low values of macro-indicators for the Philippines in table 14, which show relatively very little expenditure for R&D. In the case of copper wire manufacture, the industry is still dependent on the mother companies abroad for R&D. The same is true for coco chemicals. In these two cases the original licensing agreements required the local entrepreneur to fill his R&D needs through the mother companies. Government research agencies have been engaged in coconut research for a long time, but they have concentrated on the agricultural side of the industry. The research that has been done on the processing aspects is not relevant to the needs of industry, or, if it is, it has simply been ignored. One can say that, in the older industries, persistent colonial attitudes seem to hinder the growth of S&T self-reliance.

In the case of the newer industries (geothermal energy and semi-conductors), technological dependence is starkly apparent. The micro-indicators of interest in table 15 are the nationality of management, dependence on inputs and financing, and the importation of the hardware of production. It is interesting to note that geothermal energy is the pride of the government science establishment. Its rapid development made an impact on the local energy scene.

However, the role of Philippine science has been limited to the exploration aspects. In fact, the local expertise that has been developed in this area is of world-class quality. Sadly, however, the actual installation of machinery and pipelines was carried out by foreign multinational companies. Indeed, the Philippines is totally dependent on imported capital equipment for most of its industries.

In all the case-studies, there is evidence of heavy technology imports. Even in an old industry like mining, the biggest mine in the Philippines relies on foreign managers for the supervision of the underground operations and the maintenance of the heavy equipment. Most of the machinery is imported. In the entire copper techno-system, only the furniture is of purely local origin.[15] In the assembly and packaging of semiconductors, practically all equipment and material inputs come from abroad. Similarly, all processes and technologies are licensed by foreign firms.

As Stewart points out,[16] technology imports are addictive. In the Philippines, technology transfer is the standard way of initiating new enterprises because it appears to be the quickest and most convenient way of doing things. Once established, technology imports tend to inhibit the growth of local initiatives in the same industry because of the latter's inability to compete in terms of quality and cost. Given the lack of technological capability to develop substitutes, a local vested interest grows up around such enterprises, which end up dominating the market and assuming a prominent role in the national economy. At this stage, the continuous importation of technology becomes difficult to manage and control.

Technological dependence in the Philippines comes in many forms. In the older industries, the importation of critical inputs and capital equipment is the most common. This is followed by the actual management of enterprises by aliens, as in the case of coco chemicals and copper wire manufacture. The lack of R&D isolates these companies from the mainstream of technological innovation. Ultimately, they

Table 15. Indicators of self-reliance in science and technology: case-studies

Copper industry		Alternative energy (geothermal)	
Strong points	Weak points	Strong points	Weak points
Local maintenance of hardware	Quality of technological innovations by Filipinos	Existence of technical training programme in corporate plans	Existence of R&D programme in corporate plans
Technical training programmes in companies	Support of local R&D by the industry	Number of Filipinos with the technical know-how in relevant technologies	Nationality of management
Adequate supply of graduates	Utilization of R&D results	Number of Filipinos with managerial know-how	Number of innovations in the industry
Vertical linkages of the system	Local supply of hardware	Existence of technical industry library locally	Quality of technological innovations by Filipinos
	Number of innovations in the industry	Existence of historical industry statistics	Utilization of R&D results
	Control of financing	Utilization of locally trained technicians and engineers	Local supply of hardware
	Foreign nationals in management technological capacity	Relevance of curricula to industry	
		Adequacy of number of graduates	
		Local maintenance of hardware	
		Existence of the various components of the techno-system	
		Interdependence/linkage of the subsystem	

Coconut industry		Semiconductors industry	
Strong points	Weak points	Strong points	Weak points
Role of nationals in policy formulation	Technological capacity	Existence of technical programmes in corporate plans	Existence of R&D programmes
Nationality of management	Existence of technical training programmes in corporate plans	Number of Filipinos with technical know-how in relevant technologies	Control of technological inputs
Equity participation of nationals in corporation	Quality of technical innovations by Filipinos	Number of Filipinos with managerial know-how	Control of material inputs
Control of managerial inputs	Utilization of R&D results	Number of experts	Control of financing
Use of local material inputs to various processes	Level of R&D effort	Utilization of locally trained technicians and engineers	Innovations in the industry
Existence of technical industry library locally	Support of local R&D		Technological capacity
Existence of historical industry statistics			Existence of technical industry library
Number of scientists, engineers, and technicians in relevant fields			Local supply of hardware
Utilization of locally trained technicians and engineers			Level of R&D effort
Adequacy of the number of graduates			
Local maintenance of hardware			
Existence of the various components of the techno-system for coconut			

237

lose their competitive edge in the world markets. In the newer industries, foreign investments and financial control are the mainsprings of technological dependence; the case of the semiconductor industry is typical. Most Philippine operations are only the dispensable components of a long chain of operations leading to the marketable product. In other words, only certain portions of the techno-system are under some kind of control by Filipino nationals. In almost all cases, the learning process for Filipino technicians and engineers is of little value in the national context.

In the dependency theory of underdevelopment, it is postulated that the third world is immersed in a complex but painful array of dependent relationships with the industrialized countries. This dependent relationship spans the economic and cultural spheres. This is especially true in former colonies like the Philippines, where the values of the colonizers have been internalized and have grown deep roots. Industrialized countries' interests have developed powerful local constituencies. The most critical aspect of this entire dependent relationship could well be technological dependence. Because of the vital role of technology in the life of any nation, its control, whether direct or indirect, implies effective dominance of all the other aspects of national life. The freedom to explore alternative paths to development has been confined to very narrow limits for most third-world countries. Whether universally valid or not, the dependency theory provides a convenient conceptual frame for the understanding of the nature of technological dependence.

Since technological dependence is a multilateral relationship between a user and various suppliers of technology, governments, and international organizations, the quest for S&T self-reliance in the Philippines cannot be separated from the geopolitical context. Measures directed at S&T alone cannot succeed unless they are complemented by the influence of the technologically advanced countries. The TNCs are more affluent and powerful than many third-world countries. The economics of technology is characterized by imperfect markets in the perpetuation of technological dependence. Their awesome resources and corresponding power must be confronted with all the cunning and caution that third-world countries can muster.

S&T in the Philippines: Inputs and outputs

Reckoning from the establishment of the Bureau of Government Laboratories (now called the National Institute of Science and Tech-

nology), S&T in the Philippines was more than 50 years old when the first indictment of it was made. In a report submitted by the Chairman of the Senate Committee on Scientific Advancement, the following points were expressed:
1. Lack of coordination of research work.
2. Shortage of research funds.
3. Shortage of manpower and qualified teachers.
4. Lack of science consciousness.
In 1972, Reyes assessed the state of S&T in the Philippines in relation to other countries:

In spite of these efforts, our rate of scientific and technological progress has not been enough. A recent survey showed that the Philippines is still 40 to 60 years behind the United States; 35 to 40 years behind Russia; 30 to 40 years behind the United Kingdom, Sweden and Canada; 30 years behind West Germany and France; 20 to 25 years behind Norway and Australia; 20 years behind Poland, New Zealand and Japan. Dr Frank Co Tui, consultant to the SEATO Committee on Scientific Advancement, summed up the state of scientific and technological development in the country as "semi-primitive."[17]

Even in the 1970s the perception was that S&T in the Philippines was not being supported adequately. Reyes went on to claim that "Our expenditure in 1961 was $\frac{1}{20}$ of 1 per cent of GNP and ten years later in 1970 it was $\frac{1}{10}$ of 1 per cent."

A more quantitative assessment is possible through the use of some macro-indicators of technological capacity. Table 14 shows some of these indicators in relation to Japan and the Republic of Korea – two countries which are much more progressive than the Philippines. Table 16 exhibits a relative macro-indicator called the technology index. This is defined as the average of the sum of the number of patents and registration of new designs, technology trade, value added in manufacturing, and the export of technology-intensive goods. For inter-country comparison, the technology index for the US, the world's technology leader, is set at 100.

Table 14 suggests that in comparison to Japan and the Republic of Korea, the Philippines' serious deficiency is in what has been termed technological effort.[18] This is reflected in a shortage of scientists and engineers doing R&D and of national resources devoted to R&D. The meagre expenditure in R&D is another facet of this weak technological effort. The corresponding outcome, as measured by patents, is, as expected, of minimal economic significance.

Some quantitative indicators of the inferior position of the Philip-

239

Table 16. International comparison of technology indices, 1982 (US$ billions)

	Number of patents and registration of new designs (1)	Technology trade (2)	Value added in manufacturing (3)	Export of technology-intensive goods (4)	(1) + (2) + (3) + (4) / 4
USA	57,889 (100.0)	7.5 (100.0)	642.3 (100.0)	109.2 (100.0)	(100.0)
Japan	105,905 (182.9)	2.3 (30.9)	321.1 (50.0)	91.4 (83.7)	(86.9)
Federal Republic of Germany	16,306 (28.2)	1.4 (18.7)	244.5 (38.1)	103.5 (94.8)	(45.0)
UK	29,590 (51.1)	1.8 (24.4)	124.9 (19.4)	42.3 (38.7)	(33.4)
France	23,944 (41.4)	1.3 (17.1)	154.3 (24.0)	43.3 (39.6)	(30.5)
Republic of Korea	4,512 (7.8)	0.3 (3.7)	21.1 (3.3)	8.7 (7.9)	(5.7)
Philippines	449[a] (0.8)	−0.3[b] (0.4)	9.3[c] (1.5)	1.4[d] (1.3)	(−0.1)

Source: Korea Development Bank. For the Philippines, information on patents is from the Philippine Patent Office (figure for 1983), on technology trade from the *NEDA Statistical Yearbook*, 1985, and on value added from the World Bank, *World Development Report*, 1986 (figure for 1983).

a. Figure represents total exports of selected technology-intensive goods less total import of capital goods for 1983.

Table 17. Filipino technological capabilities

Type of technology	First-wave technologies	Second-wave technologies	Third-wave technologies
Materials technologies	Replicative in most, adaptive in some	Operative in some, adaptive in others	Pre-operative in most, operative in some
Equipment technologies	Replicative in most, innovative in some	Operative in most, adaptive in some	Pre-operative in most, adaptive in few
Energy technologies	Replicative in most, innovative in some	Adaptive in most, replicative in some	Pre-operative in most, operative in some
Information technologies	Replicative in most, innovative in some	Operative in some, adaptive in others	Pre-operative in most, adaptive in some
Life technologies	Replicative in most, innovative in some	Adaptive in some, replicative in others	Pre-operative in most, adaptive in a few

pines with respect to the industrialized countries are depicted in table 16. The most serious is the negative value of technology trade, which is also reflected by the very low value for the export of technology-intensive goods. The overall negative value of the technology index represents the stark reality of the country's technological dependence. These indicators somehow convey the "technological distance" between the countries.

A more detailed but qualitative assessment is made possible by using the S&T taxonomical matrix (table 1) and the notion of stages of technological capability. The assessment of S&T in the Philippines is shown in table 17. The evaluation was based on the results of the case-studies and the general knowledge of the Philippine situation.

A more precise definition of what it means to be an agricultural country is apparent in table 17, where it is shown that replicative and even innovative capabilities exist for all first-wave technologies. This is perhaps the result of the decades of education and research in Philippine agriculture. Unfortunately, however, agriculture cannot reach full efficiency with a weak second-wave technology. Philippine agriculture is still dependent on foreign inputs (fertilizers, pesticides, and processing technologies); there are some adaptive capabilities in equipment and information technologies, but the country is hardly in the game as far as most of the others are concerned.

Table 18. Distribution of NSTA-SPI awardees by field of study as of May 1984[a]

	Agricultural and natural sciences	Biological sciences	Medical sciences	Physical sciences	Engineering sciences	Mathe-matical sciences
Degree Programmes						
Undergraduate	64	338	–	628	802	594
Master's[b]	91	222	64	201	77	76
Doctoral	10	6	–	12	7	18
Short-term training programmes for teachers						
Summer science institutes[c]	–	–	–	–	–	–
Certification programme	–	–	–	–	–	–
Total	165	566	64	841	886	688

a. Prepared by Scientific Manpower and Institutional Development Division, Science Promotion Institute.
b. Includes graduate research fund grantees.
c. Awardees during the period 1971–1983.

Of course, Philippine S&T is not without its achievements. Appendix 2 lists the most significant accomplishments of the NSTA. Since there is very little R&D going on in the private sector, this list is indicative of the entire S&T system of the Philippines.

The large variety of research points to the lack of focus and dispersal of the already meagre funds for R&D. It is fair to say that none of these accomplishments is outstanding in the international sense. Mission-oriented R&D that is directed to specific nationally significant problems has not been addressed. In the case of geothermal energy, for instance, the lead in the exploration technology should have been carried further downstream to include the development of local capability in actual geothermal power generation. This, together with the uses of geothermal steam, could have been planned as a mission-oriented programme. The same could be said for biogas and alcohol projects.

The strong historical bias for agricultural R&D is also apparent. Industrial research has been quite inadequate. As we have pointed out, the weakness in industrial capability ultimately weakens the agricultural sector also.

The scholarship programme of the government is intended to address the weakness of S&T in respect of manpower. The result of

Social sciences	Total	Teaching						Total
		Math.	Bio.	Physics	Chem.	Gen. sci.	Total	
22	2,448	229	–	182	–	68	479	2,927
23	754	183	73	94	56	101	507	1,261
1	54	–	–	–	–	–	–	54
–		2,640	1,401	1,081	2,318	3,086	10,526	10,526
–		14	24	11	19	5	73	73
46	3,256	3,066	1,498	1,368	2,393	3,260	11,585	14,841

the programme is summarized in table 18. Although the programme was probably constrained by lack of financial resources, this kind of output will not permit the Philippines ever to catch up with the internationally set norms for manpower requirements. Technological capacity in terms of R&D manpower per 10,000 population has remained fairly static during the last few years.

Philippine R&D has no detectable impact on the national economy. This is intimated by table 3, which shows no systematic increases in the growth rates of either agriculture or industry over the years. Significant and successful innovations could have spurred growth in these sectors. The observed changes in the growth rates are perhaps the short-term effects of economic policy measures. In comparative terms, table 19 shows the performance of the Philippines and other Asian countries. The average growth rate of agriculture is comparable to that of others, including the Republic of Korea, but the Philippines has a comparatively slower growth rate in industry.

The distorted emphasis of Philippine R&D in agriculture does not show any significant effect in productivity. According to table 20, the period 1971–1978 is the lowest for the economic sectors. In other words, agricultural R&D made little difference to agricultural labour productivity. In contrast, while there has been no significant research in industrial R&D, productivity in this sector shows the largest annual growth rate. This could have been due to the introduction of imported technological innovations.

Table 19. Annual growth rates of major sectors of real GDP[a] (simple average: percentages)

	Agriculture (1971–84)	Industry (1971–84)
India[b]	1.6[c]	4.0[c]
Republic of Korea	3.6[d]	12.6[d]
Philippines	3.9	5.8
Thailand	3.9[e]	7.3[e]

Source: Key Indicators of Developing Member Countries of Asian Development Bank, April 1985.

a. Gross Domestic Product.
b. GDP data are at factor cost.
c. 1971–1982.
d. 1971–1983.
e. 1973–1984.

Table 20. Labour productivity[a]

	Labour productivity (pesos per worker)			
Year	All sectors	Agriculture	Industry	Services
1957	2,980	1,740	4,320	5,280
1971	3,620	2,370	5,170	4,640
1978	4,200	2,420	7,390	5,180
	Annual growth rates of labour productivity (percentages)			
1971–78	2,120	0.300	5.100	1.570

Source: R.L. Tidalgo and E. Esguerra, "Philippine Employment in the 1970s," PIDS Working Paper 82-02, table A-6.

a. Output per person employed is estimated by dividing national income (in millions of pesos) at 1972 prices by employment in thousands.

One major factor that could somehow explain the lacklustre performance of the R&D system is the extremely low funding levels. Although there has never been a lack of bold policy statements about the support of S&T, the realization of policy is in the actual allocation of resources. In the case of S&T, there is a wide gap between policy and practice.

Although other government agencies and the private sector are also involved in S&T activities, the budget of the NSTA adequately reflects national trends of expenditure for S&T and R&D. In the Philippines

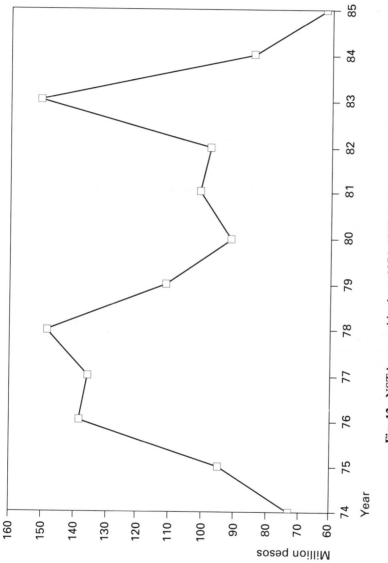

Fig. 12. NSTA general budget, 1974–1985 (index year: 1974)

the R&D expenditure of the private sector is negligible. In the case of other government agencies, the definitions "S&T" and "R&D" are very obscure and doubtful.

In general, the outlay for NSTA has been decreasing in relative and absolute terms during the last 10 years. Figure 12 gives the NSTA budget, using 1974 as the best year to correct for inflation. The sudden increase in 1983 was due the reorganization of the NSTA. Some new agencies were created and some old ones were attached to the NSTA. As shown in figures 13 and 14, there were no real increases in R&D outlay. In fact, there was a downward trend in the appropriations for R&D. These figures portray the sad reality behind the encouraging commitment of policy makers to S&T.

Table 21 shows the divergence between dream and reality. On the basis of the plan to attain a level of S&T expenditure of about 2 per cent of GDP by 1988, the annual financial requirements of NSTA were calculated. The expected annual appropriations were estimated on the basis of the historical funding increases granted by the Office of Budget Management (OBM). The last line on the table shows an ex-

Table 21. NSTA resource projections, 1984–1988

	1984	1985	1986	1987	1988
GDP (billions of pesos)[a]	461.6	530.9	610.5	702.1	807.4
S&T allocation (% of GDP)[b]	1.0	1.5	1.8	2.0	2.0
S&T allocation (billions of pesos)	4.6	8.0	11.0	14.0	16.2
Private sector share of no.3 (%)[b]	15	20	20	25	25
Government share of no.3 (%)[b]	3,918	6,368	8,784	10,530	12,113
NSTA share of no.6 (millions of pesos)[c]	823	1,337	1,845	2,211	2,544
Projected OBM allocation	683	751.3	826.4	909.0	999.9
Estimated requirements of NSTA agencies (million of pesos)	1,083.3	1,112.3	1,189.7	1,106.2	1,166.4

Source: EVSA.

a. Assumed to grow at 15 per cent annually from the 1982 level of P349 billion, at 5 per cent real growth and 10 per cent inflation.
b. Indicated in the National S&T Plan.
c. Historical average.
d. Arbitrary increase of 10 per cent annually from the amounts requested for 1984.

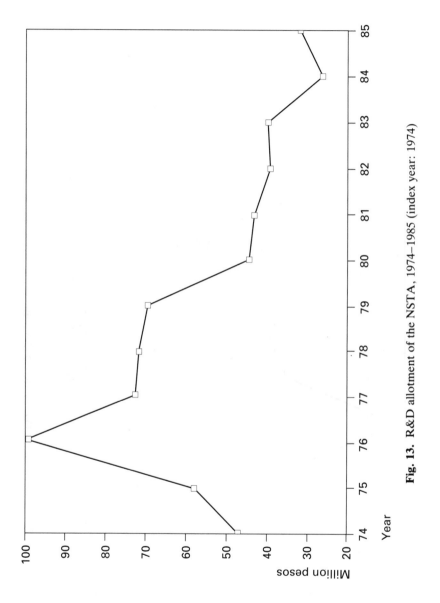

Fig. 13. R&D allotment of the NSTA, 1974–1985 (index year: 1974)

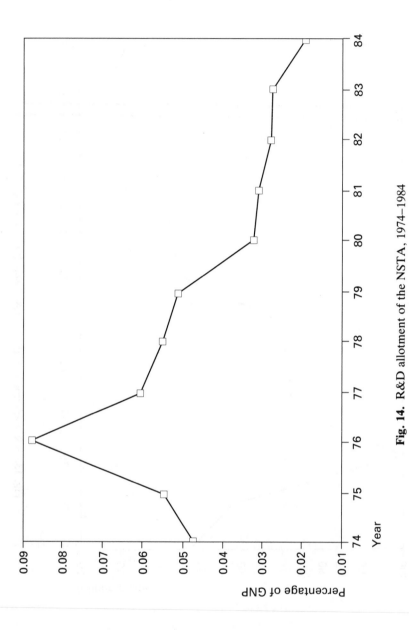

Fig. 14. R&D allotment of the NSTA, 1974–1984

pected growing shortfall. Unfortunately, the prospects for the next three years (1987–1989) are definitely much worse. In real terms, the budget of the NSTA will probably decrease.

Like most established bureaucracies in the Philippines, the NSTA has grown organizationally. Starting out as a National Science Board in 1956, it transformed itself into a National Science and Technology Authority in 1983, attaching and creating agencies in the process. The elaborate structure of NSTA is shown in figure 5. By 1987, the NSTA was once more transformed into a Department of Science and Technology. The NSTA does not have an exclusive mandate over the nation's S&T. The network depicted in figure 4 is a complex bureaucratic system which is supposed to nurture the creative enterprise of scientific and technological R&D. The situation is that there are just too many agencies, and the number is growing, making demands on a shrinking S&T pie.

It is very unlikely that things will change for the better soon. Under the new Aquino government, the usual syndrome of big words and short delivery are already apparent. The NSTA are making bold new national S&T plans, seemingly undeterred by the 30-year history of dismal funding. A recent policy paper (1986) listed an array of S&T "development strategies": there are 14 in agriculture, 7 in health, 14 in industry and energy, 5 in S&T capability and development structure, and 4 in natural hazards and environment. All these are supposed to be implemented with a budget of 93 million pesos. The document is a strange mishmash of meaningless epithets from the lexicon of previous policy exercises by NSTA. Meanwhile the inherent weakness of endogenous technology continues to worsen.

The vicious circle paradigm

After almost four decades of S&T policy formulation and planning in the Philippines there has been no qualitative improvement in the status of S&T. Certainly, there have been quantitative changes. There are now more S&T and R&D institutions, more scientists and technologists with advanced degrees, more research going on, and more laboratory equipment and tools. In the productive systems there are new indigenously developed technologies being used, especially in agriculture. Although there are new industrial facilities, these are mostly established through turnkey agreements. However, the more crucial process of what Sagasti[19] called the "endogenization" of technology has not been achieved, except in some trivial industries like soy sauce and soap manufacture. Endogenization would require a strong feedback linkage between scientific and technological R&D and the country's production systems. In general, technological skills have not gone beyond the operative stage.

The severe economic crisis of 1983 exposed the almost total dependence of the Philippine production system on the importation of the means of production and inputs. The near total inability of the local S&T system also became quite clear. The crisis of 1983 was only one of a series that have occurred periodically since the 1940s. It was just another manifestation of the underlying backwardness of S&T in the Philippines – its dependence on foreign technology and capital to sustain the economic life of the country.

To understand the anatomy of the failure of endogenization would require insights into the country's political economy and its links to the evolutionary process of growth in the national S&T culture, the nature of modern S&T itself, and the present geopolitical environment.

In the present world economic and technological order, the Philippines is at a serious disadvantage. Because of its scientific and technological backwardness, it cannot produce the equipment and machinery needed to transform raw materials into manufactured goods. As a consequence of this incapacity to produce its own means of production, the national economy has to depend on the importation of foreign technologies in the form of manufacturing processes, producer goods, and even complete production facilities in order to meet the consumer needs of the domestic market.

To finance the country's technological dependence on imported technologies, the national economy relies on the export of low value added products and raw materials such as sugar, coconut oil, logs, copper concentrates, handicrafts, and other minerals. The so-called semiconductor industries of the Philippines are merely the labour-intensive assembly operations of multinational companies. As a result, the Philippines finds itself locked into the international division of labour, playing the role of the exporter of primary commodities and importer of production technologies. It has subordinated its development to the loans and dictates of the international capitalist system.

Part of today's geopolitical reality is the growing militant awareness of third-world countries regarding sovereignty over their natural resources and economy. On the other hand, the industrialized countries are strongly asserting their proprietary rights over some vital technologies. Certainly, some hard bargaining can be expected regarding access to technologies and natural resources. Careful planning and strategy formulation will be required by third-world countries in order to obtain an equitable deal. This will require a good measure of self-reliance in S&T.

Modern S&T is very different in character from the S&T of the early years of the Industrial Revolution. Before, most industrial skills were accumulated knowledge learned through long practice. Today we have a science-driven technology which means that innovations in technology arise out of fundamental scientific R&D. There are "technology factories" controlled by big transnational corporations, where systematic mission-oriented R&D is undertaken. Some well-known examples are atomic energy, computers, and telecommunications. Commercial technologies are therefore considered to be products of a long-term investment of venture capital. These commercial technologies are the carefully guarded properties of transnational corporations. Their transfer to other parties is made with deliberate care and involves huge payments. If third-world countries are to achieve a state of excellence that can compete with the industrialized countries, they must be able to match the modern R&D infrastructures in some particular problem areas.

These are just some of the important factors that must be considered in the effort to understand why some countries, like the Philippines, seem to be trapped in backwardness and underdevelopment.

For purposes of organized analysis and strategy formulation, it is useful to construct a conceptual model to represent the salient features of the forces shaping the character of S&T in the Philippines. This model also summarizes in a concise way the rather complex feedback relationships among the numerous factors that effect the state of S&T in this country. This conceptual model is represented by figure 15, which depicts the vicious circle paradigm of S&T in the Philippines.

Some of the principal driving forces of the vicious circle are historical factors. The legacy of colonial S&T is one of the major causes of the present weakness in endogenous S&T capacity. Of course, it can be argued that there was no significant indigenous S&T before the first contacts with the West. However, the point being made here is that the colonial policies actually inhibited the emergence of a relevant and nationalistic community of scientists and technologists. During the Spanish colonial era, S&T was discouraged in favour of more classical learning. In the American era, Philippine S&T was directed towards the service of colonial objectives. Scientific and technological R&D were not linked to the local production systems. S&T was and still is not relevant to the country's economy.

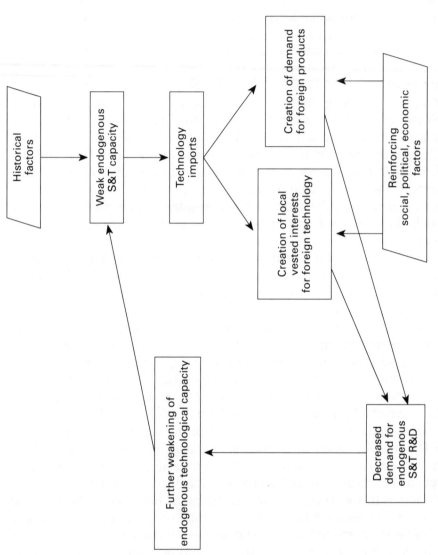

Fig. 15. The vicious cycle of technological dependence and backwardness

The momentum of these forces and the hostile social ecology of S&T that they create are responsible for the present weakness in endogenous S&T capacity. Since the local S&T is inadequate to serve economic needs, the necessary technologies are imported. The widespread use of foreign technology has many undesirable consequences: foreign investments, loss of control over decision-making and the emergence of a pattern of consumption and production based on developed-country tastes. A local vested interest in foreign technology is also created in the process. When foreign experts and executives enter a country, they easily establish strong ties with the local political and business élites. The developed country's interests are thus internalized in ways that effectively inhibit attempts to break the dependent relationships. This can strongly affect future options that are more advantageous for the host country as a whole. In this situation, local S&T becomes irrelevant to production, which further weakens it because of the lack of effective demand for local S&T products and services. The situation is a self-reinforcing, negative feedback loop that marginalizes the indigenous S&T institutions.

The other main driving forces of the vicious circle are contextual factors originating from the social, political, and economic environment. The most significant factor in Philippine social reality is the lack of self-reliant attitudes on the part of scientists and engineers. The peer group of scientists is the larger world scientific community. The local science community is not large enough to constitute a viable community with its own set of professional values. The engineering community, on the other hand, lacks a useful R&D attitude. The science group is isolated from the engineering group. There is practically no linkage between the two and both look up to the West as a model for emulation. Local scientists and engineers usually serve as consultants to foreign firms and contribute indirectly to the perpetuation of the vested interest in foreign technology and the demand for foreign products. The sense of being an identifiable and recognized actor whose views are sought and expected to influence social choices is absent.

The nature and character of the country's development philosophy implicitly abets the vicious circle. Since independence, the Philippines has deliberately courted foreign investment and technology transfer. It is obvious that the direct social costs of foreign investment are, as shown in figure 15, the stimulation of demand for foreign products and the creation of local vested interests for its perpetuation. Technology transfer is largely unregulated in terms of the actual learning process. There has been very little impact on the local S&T capacity.

The fact that the country is underdeveloped is in itself a contributory factor to the vicious circle. There are two aspects to this. One is the highly distorted distribution of wealth, in which 85 per cent of the national assets are owned by 15 per cent of the population. The result is that the wealthy have a natural preference for imported goods and the poor do not have sufficient purchasing power to encourage local production. In a sense this is another vicious circle within the vicious circle of S&T backwardness. The other aspect is the timidity of the wealthy class in risking investments in technology-intensive ventures. The outlook of the rich has always been in the traditional sectors like banking, agribusiness, insurance, real estate, and merchandising.

Some of the important political factors that tend to reinforce the vicious circle are the state's perception of the importance of S&T and the instability of the bureaucracy.

The breaking of the vicious circle will require strong political initiatives. However, S&T has not really been perceived by political leaders as crucial to the long-term success of the development programme. In spite of the political rhetoric, the problems of S&T are often overwhelmed by the more urgent political problems. It is apparent that a political consensus on the significance and priority of S&T has not yet been realized.

Since the early 1970s, the Philippine bureaucracy has been characterized by instability. Leaders at the ministerial level and the organizational structures have changed so often that it is extremely difficult to pursue a consistent policy. Even in S&T, policy directions have been in constant flux. The management of S&T has not made a dent on the vicious circle.

The anatomy of technology transfer

A general definition of technology transfer is the movement of technology into new contexts.[20] And by technology we mean the stock of knowledge required for the operation of the various components of the techno-system. In techno-system terms, technology refers principally to the information subsystem (see figure 1). When viewed in the techno-system framework, the relevance of the economic, political, and sociocultural factors in technology transfer become quite explicit. The information subsystem is physically manifested in the living minds of the social carriers of technology and various storage media. This concept is useful in clarifying the notion of the absorptive capacity of a country for technology transfer. Figure 1 also shows the linkages of the

254

information subsystem to the training and R&D subsystems. This idea enables one to understand the two broad categories of technology transfer: the transfer of commercial assets and the transfer of non-commercial assets. The transfer of non-commercial assets, which are knowledge in the public domain, usually involves the training and the R&D subsystems, and is therefore only indirectly relevant to the production activities. Most technical assistance agreements are transfers of non-commercial assets.

When the term technology transfer was first used, the meaning was restricted to the transformation of the results of R&D in the basic sciences into commercial technologies. In current usage this movement of knowledge is now called *vertical* technology transfer. However, technology transfer is now universally used to mean the movement of technology from one country to another, which is also called *horizontal* technology transfer.

Technology transfer is not a new phenomenon. Technology diffusion is a natural process. Skills and techniques are transferred from one culture to another as a result of contacts through commerce and conquest. In the past – that is, before the colonial period – the prevailing direction of technology transfer was often from East to West. Today, most of the debate in technology transfer centres on the North–South technology transfer. It must be kept in mind, however, that technology transfer between industrialized countries is of a greater magnitude.[21] The main sources of technologies at present are the US, UK, Federal Republic of Germany, and Japan. The main channels used are the transnational corporations (TNCs) which account for 80–90 per cent of technology transfers.

Technology is often transferred informally through personal contacts, readings of the literature, and professional meetings. In the techno-system framework, these could be viewed as inputs to the training and R&D subsystems and hence as not immediately crucial in productive activities.

The various formal mechanisms used in technology transfer are shown in figure 16. Direct forms of transfer include the direct purchase of capital goods and equipment, the training of nationals in specific technologies, and the hiring of foreign experts and consulting firms. The indirect mechanisms consist of the establishment of wholly owned subsidiaries of foreign companies, turnkey construction of plants and facilities, joint ventures with local companies, and variations on these dominant forms depending on the industry, national policies, and the policies of the technology suppliers. There are no established rules for

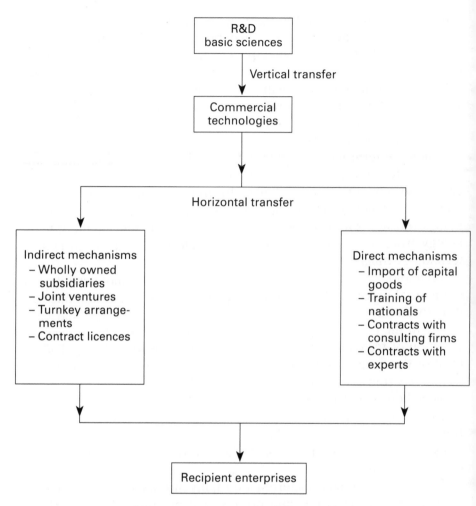

Fig. 16. The anatomy of technology transfer

obtaining the best terms. In the final analysis, technology transfer is the result of a negotiation process. The most crucial element is the ability to bargain in order to get the best terms, including the assurance that technology will really be transferred.

The developing countries must contend with the stark reality of the modern world: that the most critical resource for development – the technology of production – is controlled by a few TNCs. For example, the TNCs control over 60 per cent of the world's petrochemicals.[22] Only these few large enterprises have the necessary organization, resources, and expertise to undertake the expense and risks of modern

256

R&D for commercially competitive products. The production technologies of the TNCs are internationally tested and commercially viable. They have worldwide marketing networks.

The historical origin of the dominance of the global technology "industry" by the TNCs is traceable to the head start in empirical science by a few countries. Of the 110 significant innovations identified by the OECD in the twentieth century, 60 per cent originated from the US, 14 per cent from the UK and 11 per cent from Germany.[23]

An important aspect of technology transfer is its rapid growth in the developing countries of Asia. This is shown in table 22. Between 1972 and 1981 technology imports in the Philippines increased fourfold. In terms of technology payments for royalties and fees, the increase was sixfold. For Thailand, the increase was more than eightfold. It is important to note that even technologically advanced Japan increased its import of technology by threefold in the decade 1972–1981. Although Japan is an exporter of technology, it was still a net importer as of 1981. It should also be mentioned that there is technology transfer between developing countries; the supply of capital goods and consultancies are the prevailing mechanisms. Moreover, there is a growing number of third-world TNCs.

The experiences of Japan and the newly industrializing countries of Asia suggest that technology transfer is an essential ingredient of industrialization. The fact that modern technology is controlled by a few firms from a few countries exposes the developing countries to the dangers of monopolistic pricing, technological dependence, and inappropriate technology.

Because of the fact that technological development will certainly affect the distribution of wealth and power locally and internationally, technology transfer has a political economy dimension. In many countries, the import of technology has been associated with the emergence of a dualistic economy. One of the common features of the third world is the existence of a modern, urban, and affluent sector amidst a traditional, rural, and poor countryside. The situation is that of a micro first-world enclave. This is the inevitable result of introducing capital-intensive industries into an environment of unemployment and poverty with a feudal political economy. Unrestricted and unplanned technology transfer accentuates and perpetuates the worst features of the third world. It is not surprising, therefore, to expect that attempts to control the terms of technology transfer will be resisted by local and international vested interests.

It is quite plausible to assume that technology transfer can be a po-

257

Table 22. Payments for transfer of technology: selected ESCAP member countries, 1972–1981 ($US millions)

Country	1972	1973	1974	1975	1976	1977	1978	1979	1980	1981	Total	Three-year average, 1979–81
Developed countries												
Japan	572	715	718	712	846	1,027	1,241	1,260	1,439	1,711	10,241	1,470
Developing countries												
India	7.78	3.43	8.23	3.58	8.18	4.87	11.48	7.12	11.25	12.55	78.47	10.31
Republic of Korea	←—	96.51	—	←—	60.00	58.06	85.07	93.93	107.23	107.10	547.90	102.75
Philippines, of which:	16.76	23.56	34.01	55.73	→	63.08	62.20	63.63	72.91	67.92	519.80	68.15
Royalties/technical fees	5.81	8.83	13.62	16.56	22.01	28.51	28.56	30.74	36.32	37.59	228.55	34.88
Salaries + fees	10.95	14.73	20.39	39.17	37.99	34.57	33.64	32.89	36.59	30.33	291.25	33.27
Thailand	6.79	9.83	11.12	14.55	17.75	24.74	26.71	35.11	45.42	57.86	249.88	46.13
Developing countries combined	←——	377.81	——	——	→	150.75	185.46	199.79	236.81	245.43	1,396.05	227.34

Sources: Japan: *Science and Technology Bureau, Report on Annual Introduction of Foreign Technology*, Tokyo, 1981 (in Japanese); India: Government of India, Foreign Investment Board; Republic of Korea: Government of the Republic of Korea, Technology Transfer Centre; Philippines: Central Bank of the Philippines; Thailand: Bank of Thailand.

tent instrument to advance the foreign policy interests of a developed country. There are instances in which exports of high-technology products to the socialist block countries have been prohibited. On the other hand, the liberal technology transfer to strategically significant countries like the Republic of Korea and Turkey suggests that forces other than commercial considerations are at work. In the process of technology transfer negotiations, the developing countries would do well to take these implicit factors into account in calculating the trade-offs.

One of the imminent dangers of technology transfer is the perpetuation of technological dependence. Unless safeguards are deliberately sought by governments, the alliance between vested interests in the importing and exporting countries will constitute a powerful combination that will continue to defend and promote the existing political economy of technological dependence.

The terms of technology transfers may contain restrictive provisions which could negate the attempts of developing countries to achieve S&T self-reliance and technological mastery. The following is a summary of the onerous terms that technology transfers may involve.

1. *The cost of technology transfer.* This could be very high in monopolistic situations and where the technology is transferred in completely packaged form, as in turnkey contracts. In some cases this is accomplished through the manipulation of payments in many joint venture agreements. The favourite techniques are the overinvoicing of imports and the underinvoicing of exports.

2. *Tied inputs.* In many countries with low technological capacity, the contracts for technology transfer often contain provisions for the exclusive supply of machinery, equipment, spare parts, and other inputs.

3. *Unreasonable government guarantees.* Some exporters of technology demand guarantees against changes in taxes, tariffs, and currency exchange rates. Others ask for guaranteed remittances and royalties.

4. *Limited learning effects.* Some technology transfer arrangements are self-defeating because of the excessive use of expensive expatriate expertise when either such expertise is locally available or local people could be easily trained to the desired level of competence. Some contracts even call for the discouragement of local technological effort in the same field as the imported technology.

5. *Limits on competing technologies.* This is usually accomplished by imposing terms limiting the imports of similar technologies, or through provisions of exclusive access to local resources.

It is safe to presume that it was only during the Spanish colonization that significant technology transfer occurred in the Philippines. The principal technologies brought in by the Spaniards were those relating to construction and plantation agriculture, mainly sugar, coconut, tobacco, and hemp. The relevant manufacturing technologies that were introduced were those concerning the processing of the major crops: the milling and processing of copra and the manufacture of cigars, cigarettes, and ropes. Since this took place before the era of the commercialization of technology, needless to say, the transfers were accomplished through informal channels and by direct investments.

The technology transfer process in the early years of Philippine industrialization is exemplified by the sugar industry and the role of the Roxas family. The family dates back to the mid-eighteenth century when Juan Pablo de Roxas came to the Philippines from Acapulco. His grandson Domingo Roxas started the family's sugar business by employing a Frenchman named Gaston to experiment with sugar-cane cultivation in Batangas. While the cultivation of sugar started much earlier, the first sugar mill was constructed in 1912. Presumably, machinery and expertise were imported from Europe with gradual local adaptations. By 1930, the Roxas sugar mill had grown into a milling complex. The total Spanish investment in sugar had grown to over $20 million before the war.[24]

The history of the San Miguel Corporation, the Philippines' largest manufacturing company, typifies the transition fron informal technology transfer to the more formal, indirect mechanism of licensing agreement. The enterprise started manufacturing beer in 1880 by importing expertise from Europe, and thrived in spite of the open access of the American company, which established an office in Manila to oversee the management.

The licensing agreement has been the principle mode of technology transfer in the manufacturing of electric appliances, pharmaceuticals, transport equipment, batteries, and paints. Up to the present, all are heavily dependent on foreign technology and use foreign brands. Even in cases where local brands are used, there is a heavy dependence on foreign technology.

During the era of unregulated technology transfer in the Philippines, the dominant mechanisms used to transfer technology in support of the industrialization programme were direct investment in majority-owned subsidiaries and licensing agreements in the use of manufacturing know-how, patents, and trademarks. It has been estimated that foreign investment in the Philippines was $100 million in 1914, $300

Table 23. Foreign ownership of Filipino corporations

	Domestic	Foreign	% foreign
Food	40	21	34
Beverages	6	2	25
Tobacco	10	2	17
Textiles	29	5	15
Pulp and paper	8	2	20
Rubber	4	3	43
Chemicals	16	25	61
Petroleum	0	4	100
Non-metallic minerals	13	3	19
Metals	19	8	30
Machinery, equipment	6	7	54
Transport equipment	6	7	54
Electric appliances	5	3	38

Source: Yoshihara, 1985.

million in 1930, $315 million in 1935 and $240 million just before the outbreak of the war.[25] Table 23 shows the breakdown of ownership of domestic and foreign enterprises in the Philippines based on the first 250 largest manufacturing corporations. This table clearly shows the foreign dominance in the more modern sectors of the industry such as petroleum, chemicals, machinery, and transport equipment. Domestic companies are more active in the traditional sectors like tobacco, textiles, and beverages.

The first earnest attempt to regulate technology transfer began in 1967 with the creation of the Board of Investments (BOI) under Republic Act 5186. However, under this regulatory regime the effort was concentrated on the discrimination between pioneer and non-pioneer industries. Pioneer projects were granted liberal incentives by the government, and could be 100 per cent owned by foreigners; non-pioneer projects could only be owned by Filipinos. Projects registered with the BOI could bring into the country any number of foreign technicians.

In a study made in 1970 on foreign collaboration agreements,[26] it was found that almost 50 per cent of the sample agreements contained onerous and restrictive clauses which were unfavourable to the country. Provisions on royalties and technology were vague and non-existent, which raised the suspicion that payments were being made in other forms to evade existing foreign currency regulations. This suspicion was reinforced by the fact that most agreements were between parent companies and their subsidiaries. The initial official response to

261

this problem was a circular by the Monetary Board which limited payments of royalties to not more than 5 per cent of net sales and for not more than five years (Monetary Board Circular 393, 1973). Like most regulations promulgated under martial law, the circular provided loopholes under the guise of "exemptions" based on the "merits" of the project.

After a review of the performance of the technology transfer regulations based on the 1967 Investment Act and the Monetary Board Circular, the necessity for the creation of an institution specializing in the field was recognized by both government and industry. Thus, a Technology Transfer Board (TTB) was created under Presidential Decree 1520, which took effect in 1978. The TTB requires a prior evaluation of all technology transfer agreements before project implementation.

Rule V of the implementing rules and regulations for P.D. 1520 conveys the principal concerns of the current technology transfer regulation. This is reproduced below.

Rule V. Policy Guidelines for Evaluation

Section 1. In evaluating agreements, the Board shall be guided by policy guidelines which shall include:

(a) Appropriateness and need for the technology/industrial property right;

(b) Reasonableness of the technology payment in relation to the value of the technology to the technology recipient and the national economy as well. For this purpose, the rate of payment for contracts involving manufacturing or processing technology shall not go beyond the rate that will be established by the Board for the specific technology or industrial right to be transferred;

(c) Restrictive business clauses shall not be allowed in any agreement; specifically, the following clauses shall be prohibited:

1. Those which restrict the use of technology supplied after the expiry of the agreement (without prejudice to the application of the Philippine Patent Law).

2. Those which require payments for patents and other industrial property rights after their expiration, termination or invalidation.

3. Those which restrict the technology recipient from access to continued improvements in techniques and processes related to the technology involved during the period of the agreement even if the technology recipient is willing to make additional payments thereon.

4. Those which provide patentable improvements made by the technology recipient shall be patented in the name of the technology supplier and required to be exclusively assigned to the technology supplier; or required to be communicated to the technology supplier for its use, free of charge.

5. Those which require the technology recipient not to contest the validity of any of the patents of the technology supplier.
6. Those which restrict a non-exclusive technology recipient from obtaining patented or unpatented technology from other technology suppliers with regard to the sale or manufacture of competing products.
7. Those which require the technology recipient to purchase its raw materials, components and equipment from the technology supplier or a person designated by him (except where it could be proven that the selling price is based on international market prices or the same price that the supplier charges third parties and there are no cheaper sources of supply).
8. Those which restrict directly or indirectly the export of the products manufactured by the technology recipient under the agreement.
9. Those which limit the scope, volume of production or the sale or resale prices of the products manufactured by the technology recipient.
10. Those which limit the research activities of the technology recipient to improve the technology.

(d) The agreement shall provide that the law of the Philippines shall govern the interpretation of the contract.

(e) The agreement shall provide for a fixed term not exceeding five (5) years and shall not contain an automatic renewal clause in order to ensure adequate adaptation and absorption of technology.

Section 2. Exceptional cases. In cases where substantial benefits will accrue to the economy, such as in export-oriented ventures, labor-intensive industries, those that would promote regional dispersal of industries or which involve substantial use of raw materials, exemption from any of the above requirements may be allowed when feasible under such guidelines to be determined by the Board.

As usual, the rules and regulations contained have a deliberate loophole in section 2. There is no information available from TTB on how many applications took advantage of this provision.

After the first year of implementation, the TTB processed some 151 applications for technology transfer. Tables 24, 25, 26, 27, and 28 summarize the review by the TTB staff of the first year of operation. From these tables the following observations may be made:

1. The US is the dominant collaborator, with Japan a poor second, and the UK a very poor third. However, the US dominance has been declining. Agreements with the US constituted 67 per cent in 1970, 50 per cent in 1974, and only 46 per cent in 1980. On the other hand, Japan's role increased from 7 per cent in 1970 to 21 per cent in 1980.
2. Most of the agreements are with minority foreign capital companies

263

Table 24. National classification of agreements by type of company, 1978–1979

Country or area	Number of agreements			
	Subsidiaries/ majority foreign capital participation companies	Minority foreign capital participation companies	Purely technical collaboration agreements	Total
United States	25	22	22	69
Japan	1	20	10	31
United Kingdom	2	3	5	10
Federal Republic of Germany	2	2	2	6
Switzerland	1	3	2	6
France		3	2	5
Italy		2	3	5
Australia	1	2	1	4
Denmark	1	1		2
Sweden		1	1	2
Republic of Korea		1	1	2
Bermuda		1		1
India			1	1
Belgium			1	1
New Zealand		1		1
Panama		1		1
Netherlands	1			1
Luxembourg		1		1
Hong Kong	1			1
Total	35	64	51	150

Source: Bautista, 1980.

Table 25. National classification of licensor by products[a]

	Foods	Beverages	Textiles, clothes, etc.	Electrical supplies, appliances, and accessories	Paints and printing	Pharmaceutical materials	Metals and metal products	Petroleum products	Cosmetics, toiletries, soaps, and detergents	Motors, engines, and machinery	Cigarette and tobacco products	Office supplies and equipment	Cars, car parts, and other transport equipment	Rubber and rubber products	Paper and paper products	Telecommunications network	Plastic and plastic products	Household chemicals	Industrial chemicals	Non-metallic products	Footwear, etc.	Pyrotechnic products	Glass and glass products	Mercury pollution technology	Restaurant operation	Miscellaneous products	Vehicle-renting business	Manpower office	Data processing	Dynamic compaction	Total
United States	7		4	11	1	8	10	1	2	3	2	1	3	2	4	1	4	4	2	1			1	1			2		1		76
Japan	2		1	10	1	1	1			2		2	9	1														1			31
United Kingdom	1		2	1	1	1	1		1					1																1	10
Federal Republic of Germany						3	1																		1	1					6
Switzerland	2		1							2						1															6
France		1		1		1	1			1																					5
Italy						2	2			1																					5
Australia							2													1											3
Denmark			1																		1										2
Republic of Korea															1								1								2
Sweden																				1		1									2
Netherlands																				1						1					2
Panama							1																								1
India										1																					1
New Zealand	1																														1
Bermuda		1																													1
Hong Kong										1																					1
Belgium																				1											1
Luxembourg		1																													1
Total	13	3	9	23	3	16	19	1	3	10	2	3	12	4	5	2	4	4	2	5	1	1	2	1	1	2	2	1	1	1	156

Source: Bautista, 1980.

a. Some contracts have several product classifications included in the same contract.

Table 26. Classification of agreements by industry

Industry	Subsidiary, foreign-owned and/or controlled	Minority foreign capital participation	Purely technical collaboration	Total
Agriculture				
Artificial propagation of prawns		1		1
Manufacturing				
Foods	7	5	2	14
Beverages		1	2	3
Textiles, clothes, and accessories	1	5	3	9
Electrical supplies, appliances, and accessories (includes non-electrical counterparts)	4	11	8	23
Paints, paint materials, and printing materials		1	2	3
Pharmaceuticals	11		5	16
Metals, metal products, construction equipment and materials	3	10	6	19
Petroleum products	1			1
Cosmetics, toiletries, soaps, and detergents	1	1	1	3
Motors, engines, machinery, distribution transformers	1	5	4	10
Cigarettes and tobacco products			2	2
Office supplies and equipment		2	1	3
Cars, car parts, and other transport equipment	1	3	8	12
Rubber and rubber products	2	1	1	4

Table 27. Classification of agreements by type of assets transferred

Type of assets	Number of agreements			
	Subsidiaries/ majority foreign capital participation companies	Minority foreign capital participation companies	Purely technical collaboration agreements	Total
Patents, trademarks and know-how	15	20	9	44
Patents and trademarks		1		1
Patents and know-how		3	3	6
Trademarks and know-how	12	20	25	57
Patents			2	2
Trademarks	1	1		2
Know-how	7	20	12	39
Total	35	65	51	151

Source: Bautista, 1980.

(13 per cent), followed by technical agreements with domestic companies (34 per cent) and majority-owned subsidiaries (23 per cent).
3. Licences in electrical supplies and appliances have the biggest share
of agreements, with the US and Japan contributing equally. This is
followed by metal products and pharmaceuticals, which came mainly from the US. However, the transportation equipment sector is
dominated by Japan, with the US as a poor second.
4. The relatively high technology areas of industrial chemicals and
data processing are exclusively for the US.
5. The biggest share of the agreements (69 per cent) from table 27 involves trademarks which are mostly American and Japanese.
These empirical observations further confirm the results of the previous historical analysis concerning the colonial origins of the country's
technological dependence. The domination of technology transfer by
the US and the prevalence of trademark agreements continue to cater
to the Filipino taste for American brands (Japanese brands are now
also making headway in the Philippine market). This, of course, de-
presses the demand for local products and so inhibits the growth of
local technology. The increasing trend for purely know-how transfers
and the emergence of Japan as a supplier of technology will some-

Table 28. Classification of agreements by type of assets transferred against country of origin

Type of assets	United States	Japan	Fed. Rep. of Germany	Switzerland	France	Italy	United Kingdom	Others	Total
				Number of agreements					
Patent, trademark, know-how	21	13	2	2	1		3	2	44
Patent, trademark	1								1
Patent, know-how		3				1		2	6
Trademark, know-how	33	8	1	2	4	2	3	4	57
Patent	1		1					1	2
Trademark			1						2
Know-how	14	7	1	2		2	4	9	39
Total	70	31	6	6	5	5	10	18	151

Source: Bautista, 1980.

how dilute the historic American bias of Philippine manufacturing industries.

The industry-type classification of agreements (table 26) supports the qualitative assessment of Filipino technological capabilities represented in table 16. Moreover, the necessity for agreements in electrical equipment, which is a second-wave technology, and the absence of agreements on third-wave or high technologies further reinforce this assessment.

In their self-evaluation of the performance of the TTB, the only indicators used were the savings in foreign exchange and employment generation.

From all the available reports of the TTB, there is no information on how the contribution of technology transfer to local technological capabilities is determined. Since this is the very essence of technology transfer, it is a very significant shortcoming of the country's technology transfer regulations. Presumably, this is the responsibility of the NSTA (now called Department of Science and Technology).

The development of technological capability is a complex process. Here we can use the ideas of the stages of technological capability as a guide in determining the learning effects of a technology transfer agreement. An essential element, however, is "learning by doing." The agreements must contain an assurance that local technologists are given "hands-on" experience. In order to reach the creative stage, it has been suggested[27] that what is needed is an integrated technology transfer where know-how is blended with learning-to-know. In other words, to move forward technologically the seed must be planted, and the possibility of reaching the innovative and creative levels must not be foreclosed in any technology transfer agreement.

When a technology is transferred in a completely packaged form, as in turnkey contracts, the learning effects are minimal, and the development of each stage of technological capability will be delayed. Yet, in spite of this well-known fact, the TTB does not explicitly prohibit pure turnkey contracts.

A monitoring scheme for the purpose of determining whether technology is really transferred or transferred in the right way should be a first priority of the TTB. At the outset the details of the learning process must be spelled out clearly in the technology transfer agreements.

Conceptually and operationally, the determination of the existence of the conditions for a particular technology transfer agreement is a very difficult one. Considering the meagre resources of the TTB, it seems unlikely that this process is undertaken properly. Neither the

269

NSTA nor the Technology Resource Center, which are members of the TTB, has the necessary expertise to make an enlightened choice among the bewildering number of alternative technologies and sources. This calls for high expertise which may not be available locally. It has often been pointed out that the fundamental problem of many developing countries in this respect is one of making autonomous decisions on types, sources, and degrees of packaging of technologies. The American dominance in technology transfer in the Philippines strongly suggests that long-standing contracts and familiarity are the primary factors in the choice of technology.

Each type of technology is associated with a set of characteristics: the required inputs, the scale of production, the required supporting infrastructures, the income levels of the potential consumers, and the required skills. Those responsible for technology transfer must make sure that the transferred technology fits into existing and profitable techno-systems. In other words, the technology must be appropriate. Although the determination of appropriateness is in the rules of the TTB, the implementation is not clearly spelled out in the available records.

The usual notions associated with appropriate technology are labour intensiveness, small scale, the use of local materials, and products for low-income consumers. However, if the national objective is technological development, then it is clear that technologies promoting self-reliance and technological mastery are appropriate.

The search for models: Learning from Asia

There is merit in studying the development of S&T in other Asian countries. The combined experience of the other countries, their successes and failures, could supplement and complement our own limited experience and perhaps sharpen our responses to the challenges that we face. Japan and, to a lesser extent, the Republic of Korea loom large as possible models.

Some social observers have pointed out that seemingly crucial elements of tradition and culture that have existed in Japan and the Republic of Korea are not present in the Philippine situation. There is no such thing as a Filipino culture. Instead, we have a universe of micro-cultures with a great variety of diverse characteristics. The Llocanos are known for hard work and clannishness, while the peoples of Central Luzon exhibit their own version of the communal spirit in what is called *bayanihan*. The diversity of cultural traits and traditions of the Philippines could be an asset in this respect, and not a liability.

Compared with the gigantic problems of post-war Japan and of Korea in the 1960s, the problems facing the Philippines today appear relatively easy to tackle. Although the country is now buffeted by political and economic problems, it has some assets that could be capitalized on for technological development. The Philippines has one of the highest literacy rates in the world. As shown in table 2, enrolment in schools is comparable to that of present-day Japan and the Republic of Korea. It has a managerial class that is experienced in some second-wave technologies. It is better endowed with natural resources than Japan and Korea. More than all these, however, the world today is rife with vast technological opportunities. The third wave of civilization engendered by the twentieth century has only just begun, and there are numerous technological possibilities for "leap-frogging" into the twenty-first century. The Philippines today has more going for it than post-war Japan and Korea. The prevailing national pessimism is mostly self-perceived and imaginary. With a little dose of self-confidence and national resolve, the Philippines could catch up with the advanced countries in the early part of the next century.

Vision and commitment

If there is a commonality at all between the technological histories of Japan, China, and Korea, it is the existence of a grand vision of the future and their potential role in that future. The resources of both the government and the private sector are focused on common goals and a shared image of the future.

The Philippines is not lacking in vision. In fact, there seems to be a plethora of competing visions about the country's future. This reflects a lack of national consensus. Of course, an official long-term development plan exists, although it is not clear whether it is being supported by the present administration. In 1977, the Development Academy of the Philippines undertook a project called "Philippine Resources, Environment and the Future," which resulted in a book, *Probing the Philippines Future*. Today, this book, for whatever it is worth, has been largely forgotten by the new leaders of government. A similar interdisciplinary work entitled "The Philippines into the Twenty-first Century," under the leadership of a President of the University of the Philippines, was never published owing to lack of financial support. Recently, a big conference on an "Agenda for the Twenty-first Century" was convened by a private group. The conference degenerated into a bushfire conference and focused on the outstanding problems of the present: economic recovery, agrarian reform, delivery of justice, the role of the military, etc. The recommendations that emerged were all concerned with the immediate present. The image of a preferable or possible future for the Philippines was conspicuous by its absence. S&T was not discussed at all.

In the current Medium-term Development Plan (1987–1991), there is no apparent technology strategy. There is no reference to how the role of the Philippines is envisioned in the next century, when S&T will be the dominant world activity. Although there is a national S&T plan, this is not correlated with the planned activities of the other economic sectors. At most, there are suggestions that local S&T will be supportive of development activities but not the principal agent of growth.

In the private sector, the story is much the same. Industries, even some of the biggest ones, are not heavily involved in R&D activities. They are not motivated to innovate and they have no perception of their future competitiveness in the world of the future.

On the other hand, the S&T community has not produced anything spectacular to merit the attention of the government and the private sector.

In conclusion, the Philippine *problématique* is compounded by the lack of awareness of the value of S&T for the development process. Given this condition and the historical heritage of the Philippines, the country is trapped in a vicious circle of technological dependence and poverty. The country is besieged by political and economic problems and S&T has become buried in the turmoil of competing political and social issues.

Toward a leap-frogging strategy

The Philippines has been left behind economically and technologically by almost all its neighbours in the East Asian and South-East Asian regions during the past three decades. This suggests that the vicious cycle of S&T backwardness and economic dependence has not been broken by previous government science administrators.

With the appointment in June 1986 of a new Science Minister under a new administration and the reorganization of government science agencies under the new Department of Science and Technology (DOST) in January 1987, new science and technology policies have been formulated and new government agencies have been established. It remains to be seen whether an all-out drive to break the vicious cycle will be carried out by the new DOST; our previous analysis has shown that it takes much more than S&T policy statements and reorganization to achieve this. An entirely new national development strategy is needed to overcome technological and economic dependence.

Our initial assessment of the new government of the Philippines, however, does not evoke our optimism for national self-reliance in S&T, as the government's economic development strategies and policies do not differ much from those of the previous regime. In order to liberate the country from its economic and technological dependence, the government needs to pursue new and bold economic and technological policy directions that must attempt, first of all, to break the vicious cycle of technological backwardness and dependence.

Since the vicious cycle stems from the interrelated problems of (1) the weakness of the country's S&T potential, (2) the lack of effective demand for endogenous R&D and technological innovations, and (3) the almost total dependence of the country on the importation of technology for production, it is obvious that these three problems have to be tackled and overcome simultaneously (see figure 15).

A national strategy to break the vicious cycle is outlined in the conceptual model shown in figure 17. This includes the following essential components:

1. Accelerate massive development of the country's S&T potential through the expenditure of at least 1 per cent of GNP on the development of advanced S&T manpower, infrastructure, and information resources and the implementation of selected R&D projects.
2. Increase the effective demand for endogenous R&D technological innovations through fiscal policies and legislative acts that would make local firms, whether private or government-controlled, invest a certain percentage (at least 1 per cent) of their net income before taxes on endogenous R&D.
3. Initiate strategic management of technology transfer that would link the importation of selected foreign technologies with endogenous R&D and innovation projects for the purpose of facilitating national technological mastery of these selected technologies.

The central goal of this proposed national strategy would be the technological mastery of those selected technologies that are strategically important to the Philippine economy in its relationship with the rest of the world.

Earlier, the various technologies were categorized into first-wave, second-wave, and third-wave technologies. It was pointed out that the

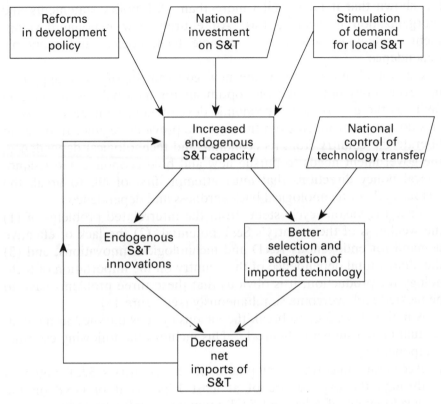

Fig. 17. Breaking the vicious cycle of technological dependence

Philippines has reached the replicative and even innovative stages of technological capability in most first-wave technologies, but that it is still largely at the operative and adaptive stages in most second-wave technologies, and at the pre-operative and operative stages in most third-wave technologies.

For the past two decades, the Philippine national debates on technological choices have been dominated by two schools of development thought: the "Countryside Development" school and the "Nationalist Industrialization" school. The former, arguing that agricultural development must precede industrialization, advocates the adoption of labour-intensive, employment-generating first-wave technologies. The latter, on the other hand, insists on following classical industrialization programmes and promoting second-wave technologies.

Under the present government, the "Countryside Development" school has gained ascendancy over the "Nationalist Industrialization" school, and the development of the agricultural sector has been included in national development plans. In fact, the "Countryside Development" philosophy was even incorporated into the new Philippine Constitution of 1987. Hence, industrialization will now be given a low priority, while agriculture-based, labour-intensive, export-oriented economic development will be pursued.

In justifying the "Countryside Development" strategy, government economic policy makers invoke the "law of comparative advantage," arguing that the country's comparative advantages lie in its abundant cheap labour, natural resource endowments, and agricultural products. Thus, they have been promoting the export of cash crops, garments, handicrafts, dolls, furniture, copra, prawns, etc.

What has been realized is that comparative advantage is not absolute and permanent but subject to technology. What has been ignored is that, today, comparative advantages are increasingly determined by scientific and technological knowledge. For example, the development of synthetic or genetically engineered products in advanced countries has determined the comparative advantages that used to be enjoyed by certain resource-rich third-world countries, while the increasing automation and robotization of production are now beginning to erode the comparative advantages of labour-intensive manufacturing in third-world countries.

In fact, three of the industries which we investigated in this project have become "sunset industries" for the Philippines because of high-technology developments. The international market for copper has been dwindling because copper wires are now being replaced by optical fibres in communication systems. New substitutes for coconut oil have been developed, resulting in the shrinkage of the export market for coconut oil. The use of fully automated systems for the fabrication of highly integrated, high-speed, sophisticated chips has started reversing the trend of setting up labour-intensive semiconductor assembly facilities in third-world countries.

What is very clear is that whatever comparative advantages the Philippines used to enjoy in the recent past owing to its natural resources or cheap labour are fast being eroded by third-wave technologies. It is also obvious that, in the twenty-first century, the economic viability of nations will be determined largely by mastery of third-wave technologies.

The current national debate between the proponents of first-wave technologies and those of second-wave technologies is, therefore, ludicrous and pathetic at a time when almost all the rest of the countries in East and South-East Asia are trying to master third-wave technologies in preparation for the twenty-first century, which is just a few years away. To pursue either a first-wave or a second-wave development strategy is to condemn the country to economic obsolescence and increased dependence.

In the face of high-technology developments that are already affecting the national economy, the Philippines can no longer afford to ignore the third-wave technologies that are radically reshaping human civilization. The only choice left for the country in the remaining years before 2000 is whether to start mastering these technologies to its economic advantage or to let high-tech development undermine its economic survival in the next century.

While third-wave technologies pose threats to the Philippine economy, they also offer opportunities for the country's economic development because of their knowledge-intensive and capital-saving characteristics. For example, the abundance of highly educated manpower in the Philippines could be turned into a comparative advantage in areas like software development, which for some time will remain a labour-intensive and skill-intensive activity. Biotechnology could also be used to lessen dependence on imported agricultural inputs and produce high-value crops, while micro-electronic instrumentation and CAD/CAM systems could be utilized to improve certain existing manufacturing processes.

The strategy of breaking the vicious cycle of S&T backwardness and economic dependence and gaining national technological mastery of selected third-wave technologies is what we propose for the Philippines. An appropriate term for this strategy is *technological leap-frogging*, because it seeks to bypass the second wave in order to (1) modernize Philippine production technologies, (2) provide a competitive edge to the national economy, and (3) bridge the technological gap between the Philippines and the advanced countries.

The essential feature of the strategy of technological leap-frogging is the linkage of selected transfers of third-wave technologies with endogenous R&D and technological innovations for the purpose of building up adaptive, replicative, innovative, and ultimately creative capabilities in these technologies.

A specific example of a technological leap-frogging approach would be the bypassing of the technology of second-wave, steel-based machine tools in favour of mastering the technology of third-wave industrial lasers for use in cutting, drilling, welding, annealing, and marking materials. The basic idea is to master, whenever feasible, the state-of-the-art technology rather than to invest money and efforts in acquiring competence in the corresponding obsolete technology.

Technology mastery refers to the innovative and creative levels of technological competence. In the example of industrial lasers, technological mastery would be indicated by the ability to improve the design and performance of existing industrial lasers or to invent and fabricate entirely new and better laser systems.

In our view, the term "technological mastery" is preferable to "technological self-reliance," because the latter is open to the misinterpretation of being either equivalent to technological autarky (i.e. the development of a technology from exclusively indigenous resources) or limited only to replicative levels of technological capability.

Besides, technological mastery connotes not only innovative and creative technological competence and state-of-the-art S&T knowledge and skills, but also the idea of "socio-economic command of technological development" – that is, the ability of the entire society to control the direction of technological innovations so as to maximize their social benefits and minimize their negative effects. In this sense, national technological mastery of third-wave technologies would imply democratic, social consensus in the selection, assimilation, development, application, and diffusion of high technologies so as to ensure a better future for everybody in the society.

The successful implementation of the strategy of technological leap-frogging, leading from the vicious cycle of S&T backwardness and dependence to national technological mastery of the third wave, is an extremely difficult national project that requires the following:

1. Strong political leadership that is fully committed on a long-term basis to this national project.
2. An effective, internationally linked national system for S&T and economic scanning, forecasting, prospective assessment, and intelligence.
3. A strong S&T system which is highly competent in adapting, replicating, and improving foreign technologies and creating new science and technology.
4. A national economic planning and management system that can formulate and implement integrated national technological and economic plans and policies in anticipation of opportunities and threats from new technological developments.
5. An economic system that is self-reliant, technologically oriented, innovative, internationally competitive, and possessed of a high degree of social equality.

6. An educational system that can anticipate and assess various probable futures and provide students with self-learning capacities for adapting to a rapidly changing society.

7. A national culture that places a high value on learning, creativity, originality, innovativeness, productivity, quality, and excellence. Furthermore, all these elements must be coordinated and integrated with one another.

In short, a radical overhaul of Philippine society is required if the strategy of technological leap-frogging is to be pursued successfully. Unfortunately, the prospects for this necessary social transformation are dim at present, for, notwithstanding the 1986 February Revolution that overthrew the Marcos dictatorship, the new Aquino government seems to lack long-term national vision for the country beyond national economic recovery.

Nevertheless, inspired by the successful experiences of Japan, the Soviet Union, China, and the Republic of Korea in technological leap-frogging, we believe that a national programme to master third-wave technologies could still be carried out in the Philippines if future national leaders could be convinced that it is imperative for national development.

Notes

1. Alvin Toffler, *The Third Wave*, New York: Bantam Books, 1980.
2. Olivia C. Caoili, *History of Science and Technology in the Philippines*, Quezon City: University of the Philippines, 1986.
3. See note 2 above.
4. Kunio Yoshihara, *Philippine Industrialization: Foreign and Domestic Capital*, Manila: Ateneo de Manila University Press, 1985.
5. UNESCO, *Manual for Surveying National Scientific and Technological Potential*, UNESCO, 1975.
6. Jose Velasco and Arcega Baens, *National Institute of Science and Technology 1901–1982. A Facet of Science Development in the Philippines*, Manila: NSTA, 1984.
7. A. Lichauco, "The IMF-World Bank and the International Economic Order," *Impact* 11(1986): 410–420.
8. Romeo Bautista and John Power and Associates, *Industrial Promotion Policies in the Philippines*, Philippine Institute of Development Studies, 1979.
9. UNESCO, *National Science Policy and Organization of Research in the Philippines*, UNESCO, 1970.
10. See note 8 above.
11. MECS and NSTA, *Science Education Development Plan*, vol. 1, Manila, 1985.
12. Ibon Facts and Figures, no. 81.
13. See note 11 above.
14. Frances Stewart, "International Technology Transfer: Issues and Policy Options," Staff Working Paper no. 344, Washington, D.C.: World Bank, 1979.

15. Pablito M. Ong, *Scientific and Technological Self-Reliance and the Copper Mining Industry*, September 1985.
16. See note 14 above.
17. Francisco Reyes, *Science and Technology in Philippine Society*, Manila: UST Publications, 1972.
18. C. Dahlman, and L. Westphal, "Technology Effort in Industrial Development – An Interpretative Survey of Recent Research," in F. Stewart and J. James, eds., *The Economics of New Technology in Developing Countries*, London: Westview Press, 1982.
19. F. Sagasti, *Technology, Planning, and Self-reliant Development: A Latin American View*, New York: Praeger Publishers, 1979.
20. F. Bradbury, ed., *Technology Transfer Practice of International Firms*, Sijthoff & Noordhoff, 1978.
21. UNIDO, *Technological Self-reliance of the Developing Countries: Towards Operational Strategies*, Development and Transfer of Technology Series, no. 15, Vienna, 1981.
22. ESCAP/UNCTC, *Costs and Conditions of Technology Transfer through Transnational Corporations*, Publication Series B, no. 3, Bangkok: ESCAP, 1986.
23. See note 14 above.
24. See note 4 above.
25. Lilia Bautista, "Transfer of Technology Regulations in the Philippines," Geneva, 1980 (UNCTAD/TT/32).
26. See note 9 above.
27. See note 20 above.

Bibliography

Bautista, Lilia. "Transfer of Technology Regulations in the Philippines." Geneva, 1980 (UNCTAD/TT/32.)
———. "Philippine Experience in Technology Transfer Regulations." Paper presented at the High-level Policy Meeting of ASEAN on the Regulation of Technology Transfer, Vienna, 28–30 September, 1981.
Bautista, Romeo, and John Power & Associates. *Industrial Promotion Policies in the Philippines*. Philippine Institute of Development Studies, 1979.
Boulding, Kenneth. *Beyond Economics: Essays on Society, Religion and Ethics*. University of Michigan Press, 1968.
Bradbury, F., ed. *Technology Transfer Practice of International Firms*. Sijthoff & Noordhoff, 1978.
Constantino, Renato. *The Philippines: A Past Revisited*, Manila. 1975.
Dahlman, C., and L. Westphal. "Technology Effort in Industrial Development – An Interpretative Survey of Recent Research." In: F. Stewart and J. James, eds. *The Economics of New Technology in Developing Countries*. London: Westview Press, 1982.
Edquist, C. "Social Carriers of Science and Technology for Development." Discussion Paper 123. Lund University, Research Policy Programme: Lund, Sweden, 1978.
Emanuel, Arghiri. *Appropriate or Underdeveloped Technology?* Paris: John Wiley & Sons, 1982.
ESCAP/UNCTC. *Costs and Conditions of Technology Transfer Through Transnational Corporations*. Publication Series B, no. 3. Bangkok: ESCAP, 1986.
EVSA Corporation. "Draft: NSTA Agency Plan (NSTA System), 1984–1988." NSTA, 1983.

Hayashi, T. "Modernization of the Copper Mining Industry." *Entrepreneurship*, vol. 4, 1982.

Ibon Data Bank. *The Philippine Coconut Industry Primer*. 1980.

Iida, K. "The Early Steel Industry: Successes and Failures." *Entrepreneurship*, vol. 4, 1982.

Industry Research Department. "The Philippine Electronics Industry: A Global Perspective." *World Bulletin* 1 (1985), no. 1.

Kusaka, K. "The Human Factor in Development: Making Liabilities into Assets." *Entrepreneurship*, vol. 1, 1982.

Lee, Chong Ouk, et al. "Self-reliance in Science and Technology for National Development of Korea." Draft, UNU Project on Self-reliance in S&T, 1986.

Maramba, Felix D., Sr. *Biogas and Waste Recycling: The Philippine Experience*. Regal Printing Company, 1978.

MECS and NSTA. *Science Education Development Plan*, vol. 1, 1985.

Ministry of Energy. *Ten-Year Energy Development Program 1980–1989*. Manila, 1985.

Needham, Joseph. *The Grand Titration: Science and Society in East and West*. Toronto: University of Toronto Press, 1969.

NSDB. "Report to the President on Accomplishments and Programs of the National Science Development Board." 1978.

Ofreneo, Rosalinda P. "Issues in the Philippine Electronics Industry: A Global Perspective." *World Bulletin* 1 (1985), no. 1.

Ongpin, Jaime V. *The Future of the Philippine Mining Industry (Or: Is There Really a Pot of Gold at the End of the Rainbow?)* Manila: Center for Research and Communication, 1982.

PCARRD. *The Coconut Industry: Problems and Recommendations*. Manila, 1985.

PDCP Industry Digest. "The Semiconductor Assembly Industry." *PDCP Industry Digest* 6 (1982), no. 6.

Quisumbing, Eduardo. "Development Science in the Philippines." *Journal of East Asiatic Studies* 6 (1957), no. 4: 127–153.

Reyes, Francisco. *Science and Technology in Philippine Society*. Manila: UST Publications, 1972.

Ramakrishna, V. "Entrepreneurship in the Japanese Context." *Entrepreneurship*, vol. 9, 1982.

Richardson, J., ed. *Integrated Technology Transfer*. London: London Books, 1979.

Sagasti, F. *Technology, Planning, and Self-reliant Development: A Latin American View*. New York: Praeger Publishers, 1979.

Silis, D., ed. *International Encyclopedia of the Social Sciences*, vol. 15. New York: Macmillan, 1978.

Soza, Mary Ann Celeste L. "The Semiconductor Industry or Foreign Exchange Earner." *Central Bank Review* 37 (1985), no. 2.

Stewart, Frances. "International Technology Transfer: Issues and Policy Options." Staff Working Paper, no. 344. Washington, D.C.: World Bank, 1979.

Stifel, Lawrence. *The Textile Industry: A Case Study of Industrial Development in the Philippines*. Ithaca, N.Y.: Cornell University Press, 1968.

Tilton, John E. *International Diffusion of Technology: The Case of Semiconductors*. Washington, D.C.: Brookings Institution, 1971.

Tobioka, Ken. "The Matrix of Nature, Culture, and Modern Technology." *Entrepreneurship*, vol. 2, 1982.

Toffler, Alvin. *The Third Wave*. New York: Bantam Books, 1980.

UCAP. *Coconut Statistics*. Manila, 1980.

UNESCO. *Manual for Surveying National Scientific and Technological Potential*. Paris, 1975.

———. *National Science Policy and Organization of Research in the Philippines*. Paris, 1970.

UNIDO. *Technological Self-reliance of the Developing Countries: Towards Operational Strategies*. Development and Transfer of Technology Series, no. 15. Vienna, 1981.

Velasco, Geronimo Z. "Geothermal Energy and Philippine Governments' Policy." International Symposium on Geothermal Energy, Hawaii, 26 August 1985.

Velasco, Jose, and Arcega Baens. *National Institute of Science and Technology 1901–1982. A Facet of Science Development in the Philippines*. Manila: NSTA, 1984.

Wang Hui-jong and Li Po-xi. "Self-reliance in Science and Technology in National Development – China as Case Study." Phase 11 Interim Report, UNU Project on Self-reliance in S&T. Tokyo: UNU, 1986.

Xu Zhaoxiang. "China's Science and Technology, Planning and Management." UNU Project on Self-reliance in S&T. Tokyo: UNU, 1986.

Yamamoto, S. "The Spirit of Japanese Capitalism." *Entrepreneurship*, vols. 1–10, 1982.

Yoshihara, Kunio. *Philippine Industrialization Foreign and Domestic Capital*. Manila: Ateneo de Manila University Press, 1985.

Appendix 1

Indicators of self-reliance in science and technology

System characteristics	Variables	Indicators	Values	Empirical references
Goal-setting	1.1 Emphasis on learning in the formulation of goals	1.11 Existence of technical training programmes in corporate plans	1	No explicit training programme in plans
			2	Training programmes not complete only for the low-skill components. Not on critical aspects of the technology
			3	Detailed training programme aimed at ultimate takeover by local experts
		1.12 Existence of R&D programme in corporate plan	1	No explicit R&D programme
			2	Some R&D programmes but on minor aspects of the technology
			3	Detailed R&D programme for product improvement
	1.2 Emphasis on local autonomy	1.21 Existence of plans for local autonomy	1	No explicit programme for the attainment of local autonomy
			2	Some aspects are programmed for local autonomy

		Score	Description
		3	Detailed programme for local autonomy, e.g. sale of equity of nationals
	1.22 Plans for vertical integration	1	No explicit plans for vertical integration
		2	There are plans for partial integration
		3	Detailed programme for vertical integration
1.3 Articulation of techno-system policies	1.31 Role of nationals in policy formulation	1	Techno-system policy fully controlled or influenced by foreigners
		2	Techno-system policies partly influenced by foreigners
		3	Techno-system policies fully controlled by Philippine nationals
2. Control	2.11 Nationality of management	1	Top-level decision makers are foreign nationals
		2	Medium-level decision makers are foreign nationals
		3	Top-level decision makers are Philippine nationals
2.1 Role of nationals in corporate decision-making	2.12 Equity of participation of nationals in corporations	1	No equity participation of Philippine nationals
		2	Partial participation by Philippine nationals

System characteristics	Variables	Indicators	Values	Empirical references
	2.2 Role of nationals in the flow of inputs		3	Equity fully owned by Philippine nationals
		2.21 Control of managerial inputs	1	Full control by foreign nationals
			2	Partial control by Philippine nationals
			3	Full control by Philippine nationals
		2.22 Control of technological inputs	1	Full control by foreign nationals
			2	Partial control by foreign/ Philippine nationals
			3	Full control by Philippine nationals
		2.23 Control of material inputs	1	Full control by foreign nationals
			2	Partial control by Philippine nationals
			3	Full control by Philippine nationals
		2.24 Control of financing	1	Full control by foreign nationals
			2	Partial control by Philippine nationals

3. Dynamics			
3.1 Innovation by Filipinos in the relevant technologies	Full control by Philippine nationals	3
3.11 Number of innovations in the industry	No important innovations at all by Filipinos	1
	Some important innovations by Filipinos	2
	Great number of innovations by Filipinos	3
3.12 Quality of technical innovations by Filipinos	No, or very little, value for the industry	1
	Some technical innovations are of high quality	2
	Most of the technical innovations are of high quality	3
3.13 Use of local material inputs to the various process	No change in the character of material inputs (or increasing use of foreign material inputs)	1
	Some increase in the use of local material inputs	2
	Most of the material inputs are now local	3
3.14 Adaptations of some of the processes to local conditions	No change in the production processes	1
	There have been some adaptations to local conditions	2

System characteristics	Variables	Indicators	Values	Empirical references
			3	There are many important adaptations to local conditions
	3.2 Change in number of Filipinos with relevant know-how	3.21 Change in number of Filipinos with the technical know-how in relevant technologies	1	No change or decrease in the number of Filipinos with technical know-how
			2	There has been a significant increase in the number of Filipinos with technical know-how
			3	There has been some increase in the number of Filipinos with technical know-how
		3.22 Change in the number of Filipinos with managerial know-how	1	No change or decrease
			2	Same but not much increase
			3	There has been a significant increase in number of Filipinos with managerial know-how
4. Systems memory	4.1 Documentation (books, manuals, journals) of the techniques	4.11 Existence of technical industry library locally	1	No specialized technical library/literature collection for the relevant technologies
			2	Same but not adequate
			3	Adequate specialized technical library

4.12	Existence of historical industry statistics	1	No statistics are being kept locally
		2	Some statistics are being gathered but there are important gaps
		3	Adequate statistics are being gathered by the local industry
4.2	Quality of information subsystem		
4.21	Technological capacity	1	Operative/adaptive capacity
		2	Replicative capacity
		3	Innovative/creative capacity
5.1	Linkage of the industry with local R&D		
5.11	Support of local R&D by industry	1	The industry/company has not supported any local R&D
		2	The industry/company has supported some local R&D
		3	The industry/company has its own R&D unit
5.12	Utilization of R&D results	1	The industry or company has not used any local R&D results
		2	The industry has utilized some of the results of local R&D
		3	The industry or company has been utilizing most of the results of local R&D

5. System feedbacks

287

System characteristics	Variables	Indicators	Values	Empirical references
	5.2 Linkage of the industry with training and educational programme	5.21 Utilization of locally trained technicians and engineers	1	Most of the technicians or engineers are foreign-trained
			2	Some of the technicians are foreign-trained
			3	Most of the technicians or engineers are locally trained
6. Systems maintenance	6.1 Adequacy of local technological educational systems	6.11 Relevance of curricula to the industry	1	Curricula not relevant to the needs of industry
			2	Curricula partly relevant to the needs of industry
			3	Curricula very relevant to the needs of industry
		6.12 Adequacy of number of graduates	1	Number of graduates is not enough
			2	Number of graduates is barely enough for the industry
		6.13 Quality of the graduates	1	Local graduates are not good enough for the industry
			2	Local graduates are barely qualified and need further training
			3	Local graduates are quite qualified to work for the industry

6.2 Adequacy of local supply of industry hardware and maintenance	6.22 Local supply of hardware	1	Almost all machinery and spare parts are imported
		2	About half of the required hardware is available locally
		3	Most of the required hardware is available locally
	6.23 Local maintenance of hardware	1	Hardware is maintained mostly by foreigners
		2	Maintenance is partly done by Filipinos
		3	Maintenance of hardware is mostly by Filipinos
7. Interdependence/ integration	7.1 Existence of the various components of the techno-system for product X	1	Only the extraction subsystem and minor subsystem exist locally
		2	Some of the critical subsystems exist locally
		3	All the critical subsystems are found locally
	7.2 Interdependence/linkage of the subsystems	1	The various subsystems are not linked with or dependent on one another
		2	Partial linkage of the various subsystems and components
		3	All the various subsystems and components are linked with one another

Appendix 2. Major achievements of S&T in the Philippines (Source: EVSA)

Several NSTA projects spearheaded exploratory efforts in new areas of research. These projects established the technical feasibility of potential technologies and provided valuable baseline data for subsequent development efforts. Several other projects were undertaken to provide scientific and technological inputs to support the government's thrusts and priorities.

Agriculture and natural resources

Development of IR20, one of the most insect- and disease-resistant rice varieties used in the Masagana 99 programme.

Development of yellow corn varieties, e.g. Protena (with higher protein content than the other varieties).

Development of new soybean varieties, e.g. Tiwala, with as much as 150 per cent yield increase over the average national yield.

Improvement of cowpea varieties, resulting in six superior varieties which outyield similar Philippine Seedboard varieties.

Development of eight disease-resistant and high-yielding mungbean varieties including Pag-asa, the first variety released by the Philippine Seedboard.

Development of high-yielding cassava varieties.

Demonstration of the feasibility of local production of wheat.

Development of local wheat varieties with commercial potential.

Improved cropping systems technology for areas that rely solely on rain for moisture.

Development of new agro-fishery techniques which allow the profitable culture in captivity of shrimps and mussels, monosex tilapia, bangus-tilapia, and rice-fish.

Development of integrated fish–crop–livestock farming system.

Monoculture and polyculture of tilapia, carp, shrimps, and bivalves.

Development of floating cage culture of fish.

Development of fish culture techniques which increased milkfish production yield from 565 to 2,000 kg/ha/yr and tilapia yield from 3 to 5 tons/ha/yr.

Improvement of breeding, raising, feeding, and management techniques of local dairy animals to develop the local dairy industry.

Production of new pig strain which performed on a par with the Yorkshire and the Hampshire from crosses between native pigs and standard breeds.

Development of a single-comb White Leghorn strain of chicken with high egg production, longevity, and hatchability.

Genetic improvements in Philippine commercial broiler chicken.

Conversion of agricultural and industrial wastes, such as banana rejects, rice hull, and straw, to animal feeds.

Substitution of costly yellow corn cassava in poultry ration.

Breeding of sunflower varieties and their utilization as a source of feedmeal and oil.

Determination of the water requirements of rice production, which formed part of the basis by which irrigation systems were designed and rehabilitated.

Innovative use of abaca for the rehabilitation of the Lake Caliraya watershed.

Adaptation of land satellite (landsat) remote-sensing as a rapid and cost-effective method of environmental and resource survey.

Improvement of plant pest and disease surveillance system, resulting in reduced field losses.

Industry and energy

Pioneering geological surveys of thermal springs as potential sources of electricity.

Demonstration of the feasibility of generating electric power using geothermal energy, so that the Philippines would become the second-largest producer in the world.

Industrial use of geothermal steam, e.g. production of iodized salt, grain drying, and fish canning.

Assessment of natural gas seepages in the country and their utilization for household cooking and lighting.

Establishment of the technical and economic feasibility of utilizing natural and man-made forests in the Philippines as sources of heat energy.

Pioneering production of biogas from animal and agricultural wastes.

Development and use of improvised strains of yeast for alcohol production.

Production of coco-diesel or petroleum fuel substitutes from coconut oil.

Development of several fuel-saving devices for automobiles which allow replacement of 20–40 per cent petroleum fuel with ethyl alcohol.

Production of producer gas from charcoal and development of its use in driving irrigation diesel pump.

Improved method of charcoal production yielding smokeless, well charred product.

Local production of charcoal briquettes.

Design and fabrication of important machines in industry, like the lumber dry kiln for the furniture industry and the abaca defibring machine.

Development of silkworm-rearing and silk-making techniques, resulting in a revival of silk production in the Philippines.

Commercial production of cotton in the Philippines.

Assessment of ceramic raw material deposits in the country to locate appropriate material for earthenware, stoneware, and refractories.

Improvement of ceramic technology, e.g. formulation of clay mix and introduction of locally fabricated equipment.

Development of clay bricks, floor and roofing tiles as housing materials.

Production of hollow blocks from soil and agri-wastes like bagasse and rice hull.

Characterization and use of secondary or lesser-known wood species.

Production of particleboards from secondary wood species and agri-wastes for use as panelling and roofing materials.

Conversion of coconut logs into lumber.

Conversion of tropical hardwood into pulp and paper, which led to the local production of newsprint.

Use of abaca fibre in preparing high-quality pulp for use in the manufacture of fine and specialty papers.

Improvement of oleoresin products from Benguet pine resins, which helped accelerate the growth of pigment and resin manufacturing industry.

Development of containers and other packaging materials for fruits and fish from indigenous materials, e.g. abaca fibre, banana stalks, coconut lumber.

Establishment of minimum thermal processes for some canned foods.

Processing and preservation of fish, meat, and vegetables.

Integrated coconut processing with products from coconut milk to activated carbon.

Production of chemicals for the cosmetic and pharmaceutical industries from coconut oil.

Creation of artificial rain through cloud seeding, using smoke generator and meterological balloon.

Determination of the meteorological parameters in the development of typhoons for more accurate prediction.

Design and fabrication of appropriate low-cost equipment for small and medium industries, e.g. chipping machines, wood-fired boilers, mixer machines.

Local manufacture of aircraft parts.

Nationwide survey of mercury and other heavy metal pollution, the results of which prompted the NPCC to impose sanctions on the industries responsible.

Support for the fabrication of prototype models.

Health and nutrition

Food consumption surveys to accumulate household nutritional data as basis for more effective policy-making and planning in agriculture and nutrition.

Development of low-cost, high-protein snack-food items, weaning food, and noodle formulations from indigenous sources like legumes and coconut.

Use of beef blood for iron supplementation.

Establishment of the relationship between aflatoxin loads and primary cancer of the liver.

Use of coconut water as a replacement fluid for children and adults suffering from diarrhoea.

Modification of approaches in health-care delivery to utilize paramedics or resident health workers.

Establishment of the prevalence, cause, and mortality record of cardiovascular diseases.

Identification and use of Philippine medicinal plants for inclusion in the Philippine National Formulary.

Pilot plant production of tablets and suspensions from such plants as *lagundi, niyog-niyogan,* and *yerba buena.*

6

Japan

Introduction

We conceive of the modernization and economic development of a nation as a continuous process involving the transformation of traditional institutions and technology by the application of modern scientific knowledge. Latecomers to this process have an advantage in being able to utilize a vast stock of scientific knowledge that has been accumulated by the early starters. The concept of self-reliance in science and technology (S&T) does not preclude making effective use of this advantage for the national attainment of these goals. A crucial issue in S&T self-reliance, in our view, is whether or not a nation can effectively adopt modern science and technology for the purposes of development without losing its identity.

Borrowing technology from abroad is one way of initiating this transformation, involving as it does the transfer of technology from the early starters. The transfer of technology, however, is a process of adoption, adaptation, and indigenization of borrowed technology. A nation's readiness, in social terms, to undergo this process is decisive. What is essential for the late starters is the ability not to imitate borrowed technology, but to create innovative imitations.

We take a broad view of technology. As we see it, it is not merely a question of machinery and equipment. As the earlier UNU "Japanese Experiences" project pointed out, technology consists of five Ms: materials to be used for the production of a given product; machines to process the materials; manpower, which combines materials and machines in production processes; management, which organizes and guides the production and marketing; and, finally, markets for the products. We think that ways to combine these five Ms organically may well lead a nation toward the attainment of self-reliance in science and technology. We have designed our research, therefore, in terms of this concept of technology.

Five stages from "technology transfer" to "self-reliance"

Progress from technology transfer to "self-reliance in technology" involves the following five stages:
1. Acquisition of proper skill and know-how.
2. Maintenance.
3. Repair (including minor improvement).
4. Design.
5. Beginning of domestic production.
After the first four steps have been passed through, the way to domestic production is open. There is a big difference between this last stage and the previous four.

These five stages are necessary for technical self-reliance, and none of them can be omitted or skipped over, though latecomers can save on time and manpower. However, as much time and manpower will be used to build up a network of related techniques and services, latecomers should not be overambitious.

It is felt that for a country to be among the leaders in modern technology, the per capita annual income should be above $10,000 for a population of at least 100 million. A longer time is required to master modern technology, for example in "oil countries" where income is high, but there is no infrastructure of knowledge.

Among developing countries, India and China are ahead of the others in this regard. The size of their resources, the number of skilled workers, the infrastructures, and their existing technical skills guarantee them this position. However, their linkages between know-how and services are as yet inadequate.

Whether or not technology transfer is possible depends on the presence of know-how and services. It is difficult, therefore, though not impossible, to transfer technology in technical fields that lack these essential prerequisites. Although it would be courageous to attempt to acquire know-how in such a field from scratch, it would not be advisable – it is more efficient to start from a "copyable" position. For a technology to be "copyable" means that the five Ms – materials (including energy), machines, manpower, management, and market (= needs) – are already in place.

Three stages to technological self-reliance

During its first stage of technological self-reliance, the level of Japan was no higher than that of developed countries in the nineteenth century. Until that time, modern machines had been made by combining various kinds of parts, for example, moulded gears and wooden or ceramic parts which were not necessarily metal-made. The collective techniques of stonemasons, ceramists, metal-workers, loom-makers and water-mill makers were utilized. A network linking national techniques ranging from the manufacture of iron and steel to that of machine tools was created, reaching down to the lowest level, and this accelerated the transfer of new techniques. Since the related techniques and services already existed, what was required was the import of simple and substantial technology at cheaper prices. This is what we call "primary technological self-reliance."

In Japan it took about 60 years to reach this stage of primary self-reliance, even though the initial conditions were favourable. In the subsequent 60 years, the present position was reached in two stages: one in which the mass-production technology of automobiles and home electric appliances was transferred and stabilized (secondary self-reliance), and the other (third stage) in which small-quantity production was diversified within the framework of mass production.

Secondary self-reliance consists in a high level of TQC (total quality control), where the culture of "Japanese-style management control" is important. The tertiary self-reliant stage symbolizes the age of automation and the unmanned factory, particularly with respect to electronics manufacture and the manufacture of production machines. Once this third stage is reached, technology development will be the main concern of management.

Degree of self-reliance of technology

The technology development of latecomers progresses from technology transfer to self-reliance. But the degree of self-reliance cannot be improved without the establishment of an overall technology capacity, especially in research and development (R&D).

Self-reliance varies according to how much R&D is invested and which technical field is focused on. A more important point is the efficiency of R&D expenditure. Although a large amount of such expenditure may bring about technological development, this cannot be regarded as useful unless it is applied for the purposes of economic development.

Figure 1 is a flow chart of the technology development of one country, showing the relation between R&D expenditure and economic development. The inputs for technology development are R&D expenditure and human resources such as scientists, researchers, and engineers.

However, as shown in the figure, R&D expenditure should be divided into three categories: (1) basic, (2) applied, and (3) commercial. A country that relies completely on imported technology at the outset will invest resources in technology development in proportion to economic development.

In an attempt to improve the degree of technological self-reliance, research on the commercialization of technology is usually conducted first. Once this is on the right track, the next steps will be taken, and R&D expenditure for applied work will be increased. At this point patents are purchased and know-how is acquired from foreign countries, but this is not enough to create a fully independent technology.

To develop technology suitable for one's own needs, much effort and financial input are required. In order to be able to produce locally what was invented abroad, one must aim at technological development that falls in the second category.

To invent things oneself and make new discoveries through basic research, a greater amount of R&D expenditure is required. To be a leader in a field, an increase in the expenditure for basic research is necessary.

Each one of the three aforementioned categories of R&D is important for total economic development. However, in the case of developing countries, the importance of expenditure should be arranged into a sequence of 3, 2 and 1, taking into account the greater efficiency of R&D expenditure. To achieve self-reliance in technology by starting from technology transfer, one needs to reverse the usual pattern of technology development shown in figure 1; that is, ones goes in the opposite direction to that shown in the flow chart. What latecomers do is to rely on imports at every stage, importing the scientific information, the patents, the know-how, the machines, and so on.

The results of research in a basic field are similar to the discovery of a scientific principle. But latecomers usually regard such results as mere information. The imported technology (patent, etc.) generally

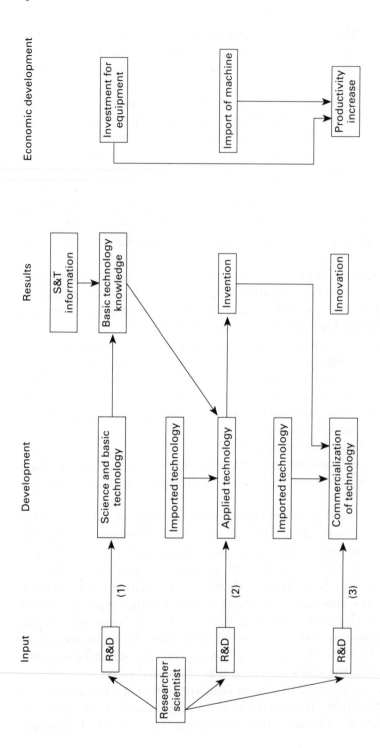

Fig. 1. Flow chart of technology development (Source: Ryuzo Sato, *Economics of Technology*, Tokyo, 1985)

Category 1: R&D expenditure for basic field
Category 2: R&D expenditure for applied field
Category 3: R&D expenditure for commercialization field

Table 1. R&D expenditure in Japan (in billions of yen)

	Category 1: basic field		Category 2: applied field		Category 3: experimental development	
	Amount	%	Amount	%	Amount	%
1975	332	14.2	505	21.5	1,509	64.3
1980	659	14.5	1,153	25.4	2,725	60.0
1984	960	13.6	1,780	25.1	4,340	61.3
USA						
1984		12.6		22.1		65.3

Source: Agency of Science and Technology, *Indicators of Science and Technology*, 1984.

has a cost, such as a royalty and licence fee, and also requires additional R&D expenditure to apply and commercialize it. Furthermore, the accumulation of research on industrialization and the operation of a small and experimental scale plant gives rise to new products, which are then distributed in the market. In the meantime, a new process is arrived at, resulting in an increase in productivity.

More than 60 per cent of Japanese R&D expenditure was, until recently, spent in the field of experimental development (table 1). This indicates that basic R&D has not yet been developed seriously in Japan, although she has been making efforts to develop her own creative technology. International comparisons, however, show that R&D expenditure in other advanced countries breaks down into similar shares for the three categories.

The index normally used to indicate how much development one country has achieved in scientific technology by her own efforts is the "ratio of R&D expenditure per national income." Figure 2 shows the international comparisons in this field. The ratio of R&D expenditure per national income is almost the same among the advanced countries, except for the USSR. In Japan, the ratio of R&D expenditure per GNP has increased rapidly and reached 2.75 per cent in 1984, one of the highest levels in the world. Taking into consideration Japan's high-growth-rate economy, R&D expenditure has increased rapidly in the past three decades.

Another index for the degree of technological self-reliance is the "ratio of technology balance of payments." Table 2 compares advanced countries' balance of payments in technology in terms of the ratio of receipts to payments. There are big differences between ad-

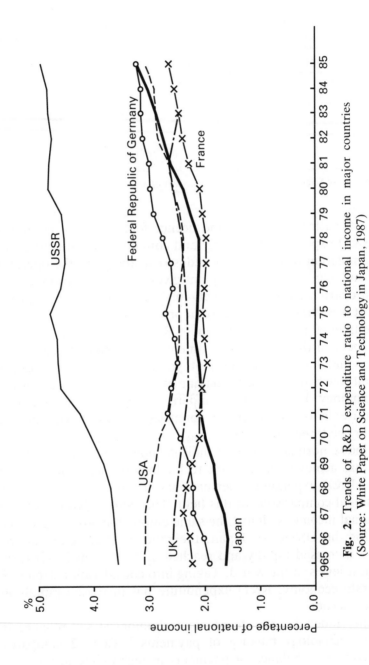

Fig. 2. Trends of R&D expenditure ratio to national income in major countries
(Source: White Paper on Science and Technology in Japan, 1987)

Table 2. Ratio of technological payments and receipts in major countries

	Japan	USA	UK	France	Fed. Rep. of Germany
1970	0.133	9.48	1.07	0.339	0.389
1980	0.267	8.68	1.24	0.483	0.486
1984	0.306	16.31	1.32[a]	0.646[a]	0.529[a]

Source: AST, *Indicators of Science and Technology*, 1985.
a. UK figure is for 1982, France and the Federal Republic of Germany for 1983.

vanced countries in their ability to export technology. Though Japan is a latecomer in this field, her ratio of the technology balance of payments has improved rapidly.

The third index of technological self-reliance is the ratio of reliance on imported technology, which is indicated by the ratio of the payments for imported technology divided by the total cost of technological knowledge (R&D expenditure + payment for imported technology).

According to an analysis made by the Institute of Investment Economics, the Development Bank of Japan,[1] Japanese industries have rapidly increased their technological self-reliance since 1970. The degree of self-reliance is calculated by dividing the technological knowledge that is accumulated through R&D by the total domestic and imported stock of technological knowledge. In this case, the Institute calculates the value of imported technology that is comparable to the domestic investment of R&D by capitalizing the payments for imported technology into their present values. The degree of dependence on imported technology was 16.7 per cent in 1970, but by 1983 it had dropped to 10.9 per cent.

A similar calculation was made by the Economic Planning Agency (EPA) in 1985.[2] Figure 3 shows the estimate for the stock of technological knowledge. In 1982, the total stock reached 17.75 trillion yen (at 1975 constant prices), which includes only 2.43 trillion yen (13.7 per cent of the total stock) of imported technology.

Low estimation of imported technology

In calculating the degree of technology self-reliance, we cannot escape from a low evaluation of imported technology. The degree of self-reliance in Japan is generally highly evaluated (fig. 3 and table 3). For example, according to a calculation made by the EPA, a simple comparison of the Japanese and US technological knowledge stock indicates that, as of 1982, Japan's annual spending reached 17.75 billion

301

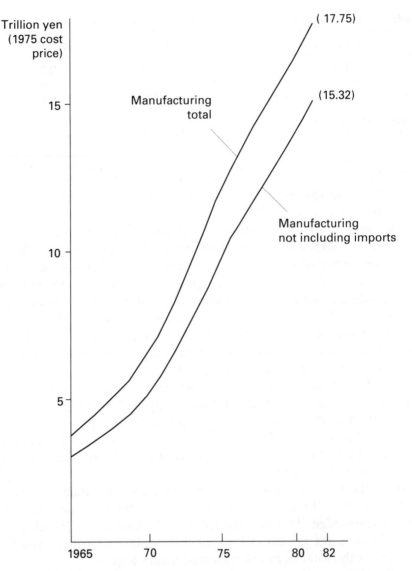

Fig. 3. Trends of stock of technological knowledge in Japan (Source: EPA, *Current Economy in Japan*, 1985)

yen (R&D expenditure spent by Japan + payments for imported technology), which is less than the one-fifth of the 86 billion yen spent by the US.

By 1982, there was no longer a technological gap between Japan and the US, especially in the manufacturing industry. For example, according to an investigation by the Industrial Science and Technology Agen-

Table 3. Degree of technological self-reliance (percentages)

Year	Flow-based (A)	Year	Stock-based (B)
1950	68	1965	–
1960	81	1970	77
1970	91	1975	80
1980	93	1982	86

$$A = \frac{R\&D}{R\&D+M}$$

$$B = \frac{R\&D \text{ stock}}{R\&D \text{ stock} + M \text{ (converted to stock)}}$$

Table 4. A comparison of the levels of key technologies (numbers of items)

	Compared with United States			Compared with Europe		
	Higher	Same	Lower	Higher	Same	Lower
Raw materials						
Development of new materials	6	2	8	7	5	2
Processing of materials	4	1	0	2	3	0
Processing and assembly						
Larger content or size	2	1	4	1	2	1
Automative and consecutive	6	6	1	7	6	2
Highly efficient production	12	18	7	18	16	2
Testing and inspection	1	2	2	0	3	2
Production control	6	3	1	5	3	0
Products						
Higher performance	13	23	22	13	28	11
Software	2	2	4	1	2	0
Design	2	2	20	6	4	10
Total	54	60	72	63	72	30

Source: Based on Industrial Science and Technology Agency, Survey Section, *An International Comparison of Japan's Industrial Technology – A Quantitative Appraisal of Principal 43 Manufacturing Sectors*, 1982.

cy, Japan was at about the same level as the US in material industries and in processing and assembling industries (table 4). Japan was slightly inferior to the US only in the fields of software and product design. As engineers' opinions were included in the investigation, the latter can be considered objective.

For the technological gap to be reflected as an industrial productivity gap, other factors such as the efficiency of capital and work practices must be considered. But the comparison reveals that the labour productivity of Japan and the US are almost at the same level.

According to an analysis by the Japan Productivity Centre, Japan's productivity in manufacturing had by 1979 already reached 83 per cent that of the US. Taking into account the difference in the growth of labour incentives in Japan and the US, it is quite possible that by 1985 Japan's labour productivity was a little higher than that of the US. It should also be noted that, in 1979, Japan's productivity was already higher than that of the US in such key areas as steel, general machinery, precision machinery, and equipment.

The technology gap between the US and Japan is, therefore, not as big as the estimated technology knowledge stock indicates. The reasons why the gap between the actual level of technology and the estimate of the technology stock is so large are as follows.

1. The US stock of technological knowledge is for military purposes, and therefore is not reflected in labour productivity. According to an investigation by the National Science Foundation, nearly 50 per cent of the total US R&D expenditure was spent on military technology (NASA included) in 1976, whereas Japan's R&D military expenditure was only 2.4 per cent of the total.
2. As Japan's investment in technology development was in applied technology and in private industries, it was effective enough to increase productivity.
3. The role of imported technology is underestimated.

Since imported technology is ready-made, products can be manufactured at low cost if manufacturing process know-how is imported with it. Many countries rely on imported technology, even in terms of paying for a licence fee, because its cost is less than that of indigenous development. In other words, importing technology is something like buying the fruits of investment that has already been made by advanced countries.

Historical perspectives on self-reliance

From imitation to creation

Japanese industrial development has relied heavily on imported technology. In the process of industrialization and modernization, Japan imported new technologies in a wide variety of fields – from brickmaking to nitrogen fixation technology in the fertilizer industry.

The development of industry in Japan over the last 120 years can be divided into four stages, as follows:
1. From the mid-1800s until the end of the nineteenth century – pure imitation of the then advanced technology.
2. From the beginning of the twentieth century to the end of the Second World War – higher industrialization with modified technology adapted to local conditions.
3. From the Second World War to the early 1970s – catching up with advanced technology.
4. From the early 1970s to the present – from imitation to creation.

Stage 1. Policies for promoting manufacturing industries by means of imported technology only

The Meiji government (from 1868 to 1912) recognized that increased production and the promotion of manufacturing industries were essential to establishing a solid economic foundation for the construction of a modern state. The immediate target of its industrial policies was the curtailment of imports of machinery, metal products, and chemicals. These poured into the domestic market with the opening of the country to foreign trade and had caused a chronic deficit in the international balance of payments. To counter this trend, the introduction of modern industry was urgently required. However, there was little private capital available, so that only direct investment by the government could accomplish the desired end.

The Ministry of Engineering, created in 1870, was charged with the responsibility for encouraging the development of many industries and running the mines, railways, and communications. The major importer of technology was the government. For example, Tomioka Spinning, located in Gumma Prefecture, was established in 1872 by the government. It was equipped with French-made spinning machines, used French techniques, and was supervised by French engineers.

In this manner, by its own example, the Meiji administration succeeded in introducing foreign industries and technology. The industrial technology of the early Meiji era, having been almost wholly dependent on technology imported from Europe and the USA, had little affinity with contemporary indigenous production techniques. This technological dependence took the form not only of employing foreign engineers and craftsmen and importing plants, machinery, and industrial raw materials, but also of importing such elementary techniques as the making of bricks. This was found necessary because of the lack of a technological tradition that could be drawn on for the development of modern industries.

S&T research agencies were at first systematically organized in the respective administrative agencies of the state. The creation of private organizations for S&T research came about rather late. From 1868 to about 1885, Japan depended entirely on foreigners for scientific and academic guidance. Thus, nearly all the teachers in the higher scientific and technological educational institutions were foreigners invited and employed by the Japanese authorities. But as time went on, they were rapidly replaced by Japanese scholars who had studied abroad or had received scientific training under foreign teachers, and who then initiated their own original courses of study. An early example of this new regime was the Earthquake Prevention Research Council, established in 1892, all of whose personnel were Japanese.

Stage 2. Self-reliance policy for science and technology

Drastic changes in the industrial structure of Japan were brought about by the policies of the Meiji government. A landmark was reached in 1919, when, for the first time, the output of industry outstripped that of agriculture. Structural changes accelerated the attainment of self-sufficiency in technology, thus consolidating the foundations of modern industries. During the First World War, when the introduction of foreign technology was abruptly suspended, the government made a determined effort to establish and realign state-run research laboratories serving manufacturing industries. At the same time, it gave friendly consideration to the views and proposals of private scientists and engineers about the opening of engineering institutes.

In an effort to devise an effective industrial policy, the government enlisted academicians and businessmen to form research councils in various fields.[3] The activities of these councils helped make the government's industrial policies truly effective. They dealt with such questions as the reduction of the price of industrial salt, the development of hydroelectric power, the promotion of technical education, and so on. They also advocated priority for the physical sciences along with the establishment of a chemical research laboratory. Impressed by their recommendations, the government established the Physiochemical Research Institute in 1917, with a government subsidy and contributions from industrial circles.

The institute was a typical example of cooperation between government and private companies. Active interchange between the different research branches was encouraged in a liberal atmosphere. The institutes earned international esteem, not only for academic performance, but also for providing business opportunities. On the technological side alone, its successful results included Masatoshi Ohkochi's piston ring, Umetaro Suzuki's synthetic *sake*, and Kotaro Honda's magnetic steel. Two Japanese Nobel Prize winners were former members of this institute's staff.

The second stage of technological development in Japan could be defined as the age of self-reliance, even though the economy still depended heavily on imported technology.

Stage 3. Process of catching up with advanced technology through further imitation

After recovering from the ruinous conditions of the post-war years, the Japanese economy achieved a 10 per cent annual growth rate until the 1973 oil crisis. Underlying these achievements were government policies promoting domestic industry. At the same time, the import of foreign technology was strongly encouraged. After 30 years, the cost of imported technology amounted to 20 times that in 1955. Thus, Japan's technological recovery owed much to imported technology. The expenditure on R&D in Japan in 1960 was only 1.2 times the cost of imported technology. The assimilation of this technology depended on a certain mature technological base that in turn was instrumental in reducing inputs, particularly those of manufactured goods, in accordance with government policies.

In the 1950s and 1960s, Japanese industry gravitated toward eventual self-reliance, thereby reducing its dependence on imports and achieving a high degree of independence in manufactured goods. The policies geared toward this self-reliance can be grouped into three categories: restricting imports of manufactured products; fostering domestic industries by protective measures; and promoting technology transfer from advanced countries.

The Ministry of International Trade and Industry (MITI) tried to foster domestic production in every manufacturing industry to increase the self-reliance of the industrial structure. Two large electric power plants, which could not be produced by Japanese electrical equipment manufacturers in the 1950s, were imported, one from Westinghouse and one from General Electric. Thereafter, Japanese contractors were forced to produce subsequent power plants through licensing and using know-how from both manufacturers.

Another example is the computer industry. In 1960, MITI decided to start a computer industry in Japan with strong protection from import restrictions. MITI persuaded Japanese users of business computers to buy Japanese brands and required the government to purchase only domestic products. Technological assistance, subsidies, the establishment of a financial company for computer leasing, and other measures were applied to nurture domestic computer production.

Japan's import-substituting policies have strongly affected the policies for S&T and caused them to deviate from an import-oriented policy. In addition to this tendency, one should note the specific technological conditions after the war. During the war, Japan's technology was isolated from foreign technologies and had specialized in the military area of the munitions industry. The technological gap in 1955 between Japan and the US was too big to be eliminated in a short time. Development through imported technology of the then new industries, such as synthetic chemicals, petrochemicals, consumer durable goods, and electronics, was urgently needed.

Whereas in 1955 more than 50 per cent of the technology imported had been developed before or during the Second World War,[4] in the 1960s Japanese industries imported advanced technology which was invented in the US after the war. So far as technology was concerned, the fundamental behaviour of Japanese big business was directed toward the imitation of foreign technology and the acquisition of technical information faster than other companies.

The imitation of technology was more effective than its creation. The importation of established foreign technology obviated commercial risk and the uncertainties inherent in the development of newly created technology. It provided a rapid and effective method of enhancing the technological level of Japanese industries.

Policies for the introduction of foreign technology
The door permitting the entry of foreign technology, which had been closed since the war years, was reopened in 1950 when the government enacted two laws dealing, respectively, with the introduction of foreign capital and with foreign exchange and trade control. These laws were designed to assist the post-war rehabilitation of the Japanese economy. Safeguards were included in the form of stipulations that such foreign technology should contribute to the improvement of the international balance of payments, and the corresponding policy on foreign exchange involved the control of the influx of foreign holders of technol-

ogy, and resulted in the selective importation of foreign technologies of such high quality that their cost in external payments was warranted. Moreover, from 1965 onward, substantial payments were received for technology exported from Japan. Of the technology imported, some 80 per cent was related to the machinery and chemical industries.[5]

As a result, production in those industries increased markedly in the latter half of the 1950s. In 1960, the value of production derived from imported technology and allocated to domestic consumption equalled total imports as calculated on the basis of customs clearance.[6] However, the export of goods produced by imported technology still remained at a comparatively low level in 1960. Japan had caught up with American standards in many spheres of technology by 1970, and was able to start selling Japanese commodities, turned out by modern industrial complexes, at relatively low prices in other countries.

Stage 4. From imitative to creative technology

According to estimations made by the Japan Productivity Centre, the labour productivity (added value base) of Japanese steel industries exceeded that of the United States in about 1973. Labour productivity in the electric appliance industries became superior in the mid-1970s, and in the automobile industries labour productivity is likely to exceed that of the United States in the 1980s.[7]

This means that technology in Japan is already on the same level as that in the United States and has surpassed that of the European nations. The total balance of payments for technology is still unfavourable to Japan. However, when the transactions are limited to newly contracted know-how and patents (table 5), Japanese technology exports have exceeded technology imports since 1977. The receipts (through export) from newly contracted patents in 1979 were almost double the payments for imported technology newly contracted in the same year.

Thus, Japan is now a net exporter of technology. According to an analysis by the Institute of Investment Economics of the Development Bank of Japan,[8] Japanese industries have reduced their dependence on imported technology since 1970. The dependence on imported technologies is calculated by dividing the accumulated intellectual knowledge of imported technology (payments for imported technology are capitalized into present values) by the total domestic and imported

Table 5. Trends in Japan's technology trade*a*

Year	Imports (billions of yen)	Exports (billions of yen)	E/I × 100 (%)
1955	11.9	0.1	0.8
1960	34.2	0.8	2.4
1965	60.1	6.1	7.8
1970	155.9	21.2	13.6
1974	159.8	57.1	35.7
1979	240.9	133.2	55.3
	(26.8)	(52.1)	(194.0)

Source: Agency for Science and Technology, Annual Report.

a. Patent fees are paid by the receiver annually in proportion to production for 20 years, until expiration of the patent rights. Japanese payment for imported technology has increased because of the increased production based on import licences. Numbers in parentheses show payments for the first year of patents newly contracted in 1979.

stock of technological knowledge. The dependence rate on imported technology was highest in 1970 at 16.7 per cent; however, in 1983 it had dropped to 10.9 per cent.

This means that Japanese industries, including high-technology and high-growth industries, have to invest more in R&D to develop their own technology, even though this is accompanied by the threat of increased risks and reduced efficiency. In response, the private sector has begun substantial investment in R&D for new technologies. In this way, Japan's government and private industry have launched a joint campaign to foster high technology, with the emphasis not only on applied research but also on basic research. Indeed, Japan is now making massive investment in R&D, equivalent to 2.5 per cent of the gross national product. This spending is exceeded only by the United States and the USSR.

Over the past 40 years, Japan has advanced from the status of a technically developing nation to that of a technically advanced nation, and has become a net exporter of technology, largely by improvements to existing technologies. Continuing as a net exporter of industrial technology might be possible in the future. However, this will require more R&D expenditure, which will tend to reduce efficiency. Japan will have to recognize that the effort to develop new technologies requires societal change rather than economic power.

According to the same analysis made by the Institute of Investment Economics, the marginal rate of return on R&D has decreased markedly from 22 per cent in 1965–1982 to 17 per cent in 1970–1982 (manufactured industries average.) The same type of analysis of US industries shows a similar tendency, 14 to 16 per cent in the 1960s and 7 to 10 per cent in the 1970s. The lower the dependency on imported technology, the higher the cost of developing new technology.

Debate continues about the degree to which technical originality or scientific creativity can be found among the Japanese, who are still widely regarded as inherently more skilful at adapting than at inventing. Japanese business is more adept in applied research than in basic research. Table 6 shows that the number of important innovations by Japanese inventors is not high, especially in terms of breakthrough inventions. However, in the 1970s Japanese inventions in the field of electronics progressed rapidly and this has dramatically increased her share of the world patent market.

Four advantages in technological self-reliance

To appreciate Japan's ability for technological development and to learn from her experiences in technological self-reliance, it is necessary to consider not only the country's national policies for S&T, but also her economic, social, and cultural background, which have enhanced technology, by and large. Four major factors which have influenced the development of Japanese independence in technology can be detected.

High capacity for the absorption of technology
Before acquiring self-reliance in technology, countries have to absorb modern technology imported from abroad. At that stage, the capacity for imitation is more important than creativity. The fundamental requirements for a capacity to absorb foreign technology are threefold: a high standard of education, assimilation of technology, and an average level of technology and scientific knowledge.

HIGH EDUCATIONAL STANDARD AT THE INCIPIENT STAGE OF INDUSTRIAL-IZATION. An increase in agricultural productivity must accompany industrialization at the incipient stage of economic development. In this connection, it is important that the educational standard of Japanese farmers was fairly high at that point. In the Tokugawa or Edo period, preceding Japan's industrial revolution, *terakoya* or temple schools (the equivalent of private elementary schools) numbered more than 15,000, exceeding the number of elementary schools established in the early Meiji era. According to an estimate made by Professor Ronald

Table 6. Number and index of major innovations in the post-war period compared with surplus measured by productivity

	Innov. listed by OECD	Innov. listed by Gellman	Productivity = GDP/workforce				Radicalness (Gellman) (%)			
			1950	1960	1970	Average	Minor	Major	Breakthrough	Total
United States	74 (100)	319 (100)	100	100	100	100	41.4	31.2	27.4	100
United Kingdom	18 (24)	85 (27)	50	50	53	51	4.4	40.0	55.6	100
Federal Republic of Germany	14 (19)	33 (10)	36	51	67	51	36.4	50.0	13.6	100
Japan	4 (5)	34 (11)	16	25	49	30	38.5	53.8	7.7	100
France	2 (3)	21 (7)	43	55	71	56	11.8	64.7	23.5	100
Canada	0 (0)	8 (3)	81	86	89	85	50.0	0.0	50.0	100
Total	112 (–)	500 (–)	–	–	–	–	43.7	36.7	28.6	100

Source: H. Inhaber and M. Alvo, "World Science as an Input-Output System," *Scientometrics* 1 (1978): 43–46.

P. Dore, the male literacy rate toward the end of the Tokugawa period was 40 per cent, while the female rate was in the order of 19 per cent.⁹ These percentages are much higher than the literacy rate today in many developing countries in the early stages of industrialization.

Japan's high literacy rate at the time made it easy to secure a high-quality labour force for the subsequent process of industrialization. It also facilitated the introduction of new strains of crops into farming villages. The introduction and expansion of elementary education from the beginning of the Meiji era contributed to the subsequent economic development. However, it has to be borne in mind that Japan had enjoyed a high educational standard prior to industrialization.

ENTREPRENEURIAL SPIRIT AND STRONG WILLINGNESS TO LEARN NEW TECHNOLOGY. Japanese enterprises have shown a strong propensity for new technology. A notable characteristic of Japanese technical progress is the rapid diffusion of new technology in Japan. No new technology can help to strengthen national competitiveness unless it is applied to the production lines of many companies.

An analysis by Professor Edwin Mansfield shows that conversion of half the steelmaking capacity in the United States to basic oxygen furnaces took 13 years after the development of that technology.¹⁰ This reflects neglect in replacing existing furnaces with new ones, regardless of the fact that basic oxygen steelmaking technology opened the way to economical production of high-quality steels. In Japan, basic oxygen furnaces accounted for 60 per cent of total capacity only seven years after the introduction of the American technology. This example suggests the reason why Japan has overtaken the United States in steel-making, expanding its productivity to 1.5 times that of the US. The lesson is that in addition to the development of new technologies, the readiness to abandon old technologies, even with risk, is essential for the successful imitation of technology.

Another characteristic of the Japanese ability to absorb technology is the ingenuity of Japanese workers in responding to technological innovation. They feel a sense of intimacy with technology. American and European workers have often opposed the introduction of new technology, including the use of robots, but Japanese workers rarely think in this way. In Japan's automobile factories, for example, each robot is called by a pet name. Robots are regarded as friends who can take over menial, unwanted chores; robotization in Japan is thus steadily expanding. To American workers, on the other hand, robots are only enemies that threaten them with possible unemployment.

HIGH LEVEL OF TECHNOLOGY IN SMALL-SCALE INDUSTRIES. While considering the capacity to absorb technology, we cannot neglect the technological gap among domestic industries. Japan's technological diffusion was accelerated by her specific industrial structure. Japan excels in technology where mass production is an important element, and where, at the same time, shared precision machining and assembly are prominent. Japanese competitiveness in the world market is strong in specific manufacturing industries such as automobiles, TVs, cameras, and other electronics. Generally, these industries rely heavily on parts makers and subcontractors.

The Japanese industrial structure is characterized as a dual economy in which large- and small-scale industries coexist harmoniously. Parent companies, which produce automobiles and TVs, buy and assemble their parts. To maintain high quality and high productivity in the final products, they have to utilize the same quality-control systems and highly organized delivery systems as the subcontractors.

The modern technologies introduced in big business at the beginning have been rapidly disseminated to small-scale factories, which are supported by the parent company not only in technological know-how but also in management and finance, under the so-called pyramid industrial structure. Furthermore, the Japanese industrial structure, which was historically developed and modernized over a long period, accelerates the closing of the technology gap between big business and small-scale factories.

Narrow technological gap at the beginning of industrialization
Although Japan had lagged behind in industrialization, her technological standards in the development of the heavy and chemical industries at the beginning of the 1930s were not far behind those of the Western advanced countries. For example, the production of ammonium sulphate began in the 1930s using a nitrogen fixation method. Although the main technology for this had been imported from the West, Japan succeeded in industrializing her own technology, which had been domestically developed by the Government Experimental Station (Tokyo Kogyo Shikenjo.) This means that the Japanese ammonium sulphate manufacturing industry had secured a technological standard high enough to compete with foreign technology. Also, in the post-war period, the Tore company, one of the biggest producers of rayon fibre, started manufacturing nylon by introducing nylon technology. Although the company had already developed the tech-

nology at that time, because of patent considerations it had to introduce it from the inventor, Du Pont (USA).

Thus, the high standard of basic technology had made it possible for Japan to catch up with the Western advanced countries. In the developing countries today, the mere introduction of technology will not lead to success in industrialization because the gap in large-scale, sophisticated technology between the more and the less developed countries is too wide.

Non-reliance on foreign capital in the process of capital accumulation
As a result of the economic development process characterized by the high saving and investment propensities, Japan did not attempt to accelerate her industrialization by inducing foreign capital, though she temporarily borrowed money from abroad. This is partly because the Japanese have a peculiarly negative feeling toward foreign capital. With the exception of the petroleum-refining industry, Japan introduced technology but did not induce capital. Brazil, for instance, relies on foreign capital for more than 5 per cent of its annual net investment in capital equipment. In contrast, Japan hardly depends on foreign capital at all.

Many of the developing countries in Asia either restrict the induction of foreign capital or treat domestic capital preferentially in order to avoid either extreme dependence on foreign capital or being ruled by multinational corporations. In their attempts to establish basic industries, however, dependence on both foreign capital and technology is normally unavoidable. This has brought about such problems as a failure to develop the national economy or to improve the people's welfare, in spite of industrialization. On top of these problems, MNEs (multinational enterprises) prefer to make decisions on technology to suit their own interests. They do not care for the development of local technology or the modification of traditional technology. Self-reliance in technology should be achieved through self-determination in technology.

The dual structure of Japanese culture
One of the reasons why industrialization and modernization progressed rapidly in Japan is that the Japanese have accepted foreign culture without resistance. Earlier, the Japanese had established a unique system by accepting Chinese culture. Japanese culture, therefore, has been "mixed" from the very beginning. It is due to this high

315

adaptability to alien culture that, during the modernization process when industrialization was beginning, some people proposed that the English language supersede Japanese. However, it should be specifically pointed out that, despite this openness to acculturation, everyday life in Japan does not change readily. The dual structure of everyday life, as exemplified by some Japanese wearing business suits at work and changing into the traditional *dotera* or padded dressing gowns at home, can be said to have expedited adaptation to alien culture.

Japan has attached great importance to science and technology throughout the history of her modernization. However, the leader of Japan's modernization, Sakuma Shozan (1811–1864), at the end of the Tokugawa era, advocated the famous slogan "Japanese Morals and Western Arts." Shozan thought that the combination of local ethics and foreign technology was the key factor in modernization. In fact, after the Meiji Restoration, Japan accepted Western science and technology without reserve, while she recognized the value of her independence in the realms of philosophy, morals, and culture itself.

Another example of this dual cultural structure may be that, despite the recent, total Westernization of life style, the Japanese make a sharp distinction between foreigners and their fellow countrymen in terms of ways of thinking or interpersonal relationship patterns. The fact that English words are distinctly distinguished from Japanese words in writing by the use of *katakana*, or the square form of the Japanese alphabet (though many English words have been Japanized and many 'Japanglish' words have been coined), serves as yet another example.

The dual cultural structure may be said to have expedited the modernization of Japan without impairing the traditional value systems. Japan was thus able to establish her own technology and to promote policies for self-reliance in science and technology.

Case-studies

The general problems related to Japan's acquisition of technology are best illustrated by the case-studies of two technologies, the food-processing and the electronics industry. The first is an industry initially embedded in tradition, in that many of the foods were known and made in pre-industrial Japan. The second industry, electronics, is today synonymous with the highest technology, in which Japan is now a world leader. These two case-studies, with their judicious mixture of tradition and modernity, provide an insight into the Japanese mastery of technology.

The food-processing industry

Grafting a modern technology onto a traditional one

Agro-based industry links the agricultural with the non-agricultural sector. The development of this industry, therefore, depends on the simultaneous growth of these two sectors, and in this regard it may well reflect a pattern or patterns of national economic development.

The food-processing industry depends to a significant extent on related industries such as the storage and distribution of processing materials and processed outputs. Also, it has to be supported by the manufacturing industries which produce processing machinery and equipment, materials used for packing, wrapping, and filling, and transport machinery for distribution. Thus, these related industries are integrated within the framework of the food industry.

Every country has a variety of traditional food industries based on well-established indigenous processing technologies. These traditional technologies have been improved by the application of modern scientific knowledge. In this sense, there is clearly a process by which traditional technologies are combined with modern ones, creating a hybrid of the two. We believe that the transfer of technology is a widening cycle of adoption, adaptation, and indigenization of borrowed technology. The creation of hybrid technology must be an essential part of the cycle. The food industry, with its extensive traditional technology, provides an excellent example of this hybridization process.

The interactions of five Ms, the extent and nature of inter-industry technological linkages, and the diffusion process and mechanism of technological information in the food industry were given high priority in the research. Furthermore, R&D activities within a firm, in the private sector, and in the public sector are carefully studied.

Using the detailed case-study, we can divide the technological development of the Japanese food industry into three stages.

The first stage, up to the Second World War, was a period of stagnation, except for the development of the canning industry. Most manufacturing was carried out by small producers or shops, and the food habits of the Japanese centred around fresh foods. The only other demand was for salted and dried fish and seasonings. Flour milling, sugar refining and food oil manufacture were the only food-processing industries in this period.

317

The second stage was a "leaping" period. Westernization, and diversification of the daily life of the Japanese, dramatically increased the demand for processed foods, and brought in mass production. The modernization of the bread-baking, the dairy products, and the meat processes were now developed. There was also the development of the eating-out industry and a rapid increase in the consumption of processed and instant foods. The increase in consumers' income and the penetration of electric cookers and electric refrigerators were major factors in the changes in consumption. The rapid expansion of demand for processed food led to the modernization and the quick growth of the industry. The market developed enough to warrant large-scale production. During this period, technology development concentrated on increases in productivity and the introduction of mass production.

In the third stage, technology improvements in related industries contributed to the development of the food industry. Things took a dramatic new turn in such areas as the development of packing machines and packing materials (plastics, etc.) and the computerization of freezing devices.

The import and improvement of overseas technology was the initial stage. Once imported technologies had been well assimilated and digested, they were then improved to conform to the economic and social conditions of Japan, and were further developed and "Japanized." Canned tomatoes were one example. Techniques such as canning and tomato cultivation and the know-how of contract cultivation, which had been imported from the US, were improved. Plant breeding to produce species suited to the natural conditions of Japan, and other improvements of technology, were achieved through trial and error. The industry made much use of the "experiences and perception" of workers on the shopfloor. The purpose of imitation was not just to make a dead copy, but, through a learning process, to exploit fully these accumulated "experiences and perceptions."

Although the Japanese food-processing industry had modernized itself mainly with imported technology, constant efforts were also made to achieve self-reliance. The efforts of the National Food Research Institute towards the technology development of the food-processing industry were crowned with great success. The R&D efforts of local agricultural experimental institutes and food industry institutes were also beneficial in terms of the specific conditions of individual districts. Since local features strongly influenced the food-processing industry, many skilful technology leaders appeared throughout the nation, even at small research facilities, and made a considerable contribution to the modernization of the Japanese food industry.

In recent years, the food industry has applied the achievements of gene technology to the food-processing industry, and has moved successfully into a new field of production.

In particular, monosodium l-glutamate monohydrate (MSG, called *aji-no-moto* in Japanese), which has become a huge enterprise, is an outstanding example. This product, invented in Japan in the 1900s, was a rare case of independent invention. The Ajinomoto Company is a leader in food-processing technology. All further development of the technology, ranging from adoption of fermenting techniques to the development of large-scale synthesis methods from oil, was carried out by the company's own research institutes. This is a good example of technological self-reliance, in which continuous efforts to achieve technological development created new manufacturing methods.

In general, the high degree of self-reliance was the result of much effort combined with (a) fundamental research by universities, (b) applied research by the state research organization, and (c) the R&D expenditure of the research institutes of private enterprises.

The technological development of related industries was also important for the improvement of the food-processing industry. Improvement in the technology of the producers of raw material (farmers) plays an important role. The correlations between the canning industry and the technology of horticulture were also important. An additional contributing factor was the standardization of agricultural products by the government through the establishment of JAS (Japan Agricultural Products Standard). The technological development of food-processing machines and improvements in packing materials and packing machines were significant and should be viewed as innovations in marketing, one of the five Ms. Taking soy sauce as an example, changes in packing materials from barrels to bottles, and then to plastic bottles, also prompted innovations in the area of transportation.

Modern and traditional technologies

It is significant that the traditional food industry already had a high level of processing technology in the middle of the Meiji era, when industrialization began. Fermenting technology had also been developed for the production of *miso*, soy sauce, and *sake*. The skilful use of these traditional technologies, the introduction of scientific methods, the adoption of processing methods suitable for mass production, and the development of marketing, which brought about a big rise in consumption, were the basis for the remarkable development of Japanese food-processing technology.

Taste and flavour are characteristics not only of foods, but also of associated bacteria and fungi. These differ from one district to another, and are also subtly influenced by climate and other natural conditions. This is where traditional technology can be utilized. Since the food habits of a country depend on its culture, traditional processing methods cannot be radically changed by the adoption of scientific technology alone.

As the taste and flavour of soy sauce have played an important role in the traditional food habits of the Japanese, one cannot dispense with the yeast, even for the purposes of industrialization. The brewing process was speeded up by replacing the soybean protein with amino acid, after researchers had studied the biology and biochemistry of the necessary microbes. The development of research on the microbes increased the earning ratio of amino acid, and succeeded in shortening the brewing time. But the traditional skills of experienced workers were utilized here, and were an integral part of the improvement.

Technological development based on a grass-roots movement

The Japanese food-processing industry consists of a mixture of small and very large enterprises. Technological innovations are carried out by the huge enterprises, but an important characteristic of Japanese industry is that minor enterprises are aggressive enough to absorb the new technology and are very curious about it. The field survey has proved that the minor enterprises have an excellent entrepreneurial spirit and do not hesitate to invest in technology improvement.

What is particularly significant here is that government support and guidance to minor enterprises also contributed to good results. Additionally, as in the case of soy sauce, the technology developed by huge enterprises was made accessible to other enterprises free of charge for the sake of overall industrial development. This also contributed to a levelling up of the technology standards of minor enterprises. Another contributing factor was the establishment of a canning technology training centre by private enterprises (tin-makers) at their own expense.

The special nature of Japanese society, in which "conciliation and competition" coexist, has given the technological development of the food-processing industry some of the characteristics of a grass-roots movement. Moreover, the role of the Agricultural Cooperative Union in the development of the industry should not be ignored. This development, which has also embraced the farming community, has resulted in the spread of technology and the increase in the number of creative ideas and inventions coming from the shopfloor.

The electronics industry

In addition to the food industry, the Japanese team has done new empirical studies on the electronics industries, especially the semiconductor industry. These industries differ from the food industry in that they can only be developed through their own creative technologies. Our research, therefore, was conducted in the following sequence.

1. The identification of technological structures and the linkages between these industries and related ones.
2. An analysis of the process of indigenization of borrowed technology in the electronics industries and of government policies for self-reliance.
3. The organizational structure of these industries, with a special emphasis on R&D expenditure practices.
4. Technology transfer by Japanese multinational enterprises through direct investment abroad.

The theoretical framework for the analysis of the semiconductor industry embraces three systems: the technology system, the organization system, and the inter-organization systems. Why Japanese corporations became successful in the semiconductor industry, although they started out by imitating US technology, was very much a question of their flexibility of management strategy. Our research concentrates on the relationship between technological development, companies' strategies, and government policies.

The case-study describes how and why Japanese integrated circuit (IC) technology has outrun that of the US and achieved self-reliance in high technology.

Keen competition in the consumer product market

One of the reasons why Japanese corporations became successful in the IC industry was their choice of product market. They produced semiconductors and ICs for commercial goods rather than for military and space use. Keen competition in the market stimulated drives for product innovation, such as in transistor radios, portable colour televisions, and calculators, and for production efficiency, in order to produce cheaper, better quality goods. It was therefore inevitable that Japanese corporations would gradually increase their R&D expenditure in order to keep up with the newest technological developments and to develop an indigenous, self-reliant technology that would cut the cost of patents and technical agreements and give them the leadership in the market.

In the period of innovative imitation of the transistor stage, research activities centred on the utilization of already existing semiconductors to produce new commercial goods. This was market-oriented innovation. However, with the accumulation of basic knowledge and the constant effort to catch up with the latest technologies, the emphasis began to shift from market-driven to technology-driven innovation. Competence in technological innovation became the key to success at this period, and greater efforts and larger expenditures for research activities were therefore required. It was this successful transformation from market-competition-driven innovation to technology-driven innovation that enabled Japan to obtain the world leadership in the field of memory IC. Thus, choice of the market and R&D-oriented corporate strategies did much to stimulate the development of the technological system.

Flexible organization in tune with the developmental stage of technology
The second reason for success was that even big Japanese IC companies avoided bureaucratic organizational structures and used flexible organization to support technology development. There was evidence of a fairly orderly development of the R&D organizational structure. As the demand for innovativeness increased, the layer of research activities was increased from one to three steps: (1) basic research in the central research institutes; (2) applied research for both technological development and commercial product development in the technical centre outside the division (but inside the division group); and (3) immediate technological and commercial research within the division.

Furthermore, in order to bring about changes in the social system without being restricted by the current bureaucratic organizational structure and to create the proper environment for researchers, task forces and outside ventures were and are made use of. Such clear correspondence between the innovativeness of the technological system and the changes in the R&D organizational structure indicates a need for the corresponding development of the organization in order to promote development in the technological system.

Coordination of business and government
The technological development of Japanese IC industry has often been thought to be heavily supported by the government. However, it should be pointed out that the coordination of private enterprises and the government is a strong factor contributing to the realization of self-reliance in technology.

Organizational interdependency also contributes a great deal to the development of the technological system. In the case of Japan particularly, the government was the most influential party in stimulating development. It reduced uncertainty, not only by providing protection but also by stimulating research activities. Government and business relationships with regard to the development of the technological system could be described as having a reverse U shape.

In the initial period, government research institutes took the initiative in promoting research activities in the transistor field, and helped corporate researchers by providing information, grants, and targets. When it came to the second period of the transistor and IC stage, the government began to pursue the two policies of protection and promotion of research activities. In the third stage, the policies of liberalization of some IC industries were added in order to cope with trade pressures from the US, but the protection and promotion policies remained the same.

As a result of these policies, Japanese corporations came to possess sufficient technological capabilities to compete against the US. Thus, in the VLSI stage, government policy began to concentrate on the establishment of a background support system in order to promote research activities. Thus, the characteristics of interorganizational relationships were transformed along with the changes in the technological system.

Effects of infrastructure of science and technology
The infrastructure of science and technology has to be regarded as a more important interorganizational relation. In particular, the utilization of manpower that has received a thorough education in high-technology industry plays an important role in the establishment of self-reliance.

Future of IC technologies and industries
In IC technological development, the progress of integration will never see an end; it will progress to a further stage, even after the "mega-bit age." We are now in the mega-bit age of this super integration. In

Japan, production of the 1M and 4M bit class is already under way. The range above 16M bit class will be called ULSI (U for ultra).

In Japan, the manufacturing technology for VLSI was developed over a short period. Success was achieved by the development project which brought in the Super LSI Research Technology Association, a semiconductor industry association outside the framework of private enterprise and under government guidance. These results therefore drew attention both at home and abroad. The objective of the project was to develop the necessary elements for the development of a computer of the next generation.

The application of VLSI created the personal computer, smaller than the mini-computer, and was then popularized. During the same period, after the personal computer, the word-processor was developed as a response to the long-standing need for a Japanese typewriter that was as convenient as an English typewriter.

This trend in O.A. (office automation), F.A. (factory automation), and H.A. (home automation) is now the leading factor in high technology. The application of these semiconductor products constitutes the "micro-electronics revolution," which at present is having a social impact in many fields. In the pursuit of goods that are smaller in size, lighter in weight, longer in life, and more reliable, the semiconductor industry is now not only looking forward to the age of a mega-bit class using silicon material, but also laying stress on the development of semiconductors made from a combination of gallium and arsenic. There will be a period during which both will coexist.

Japan's semiconductor industry achieved an output of 2.5 trillion yen in 1984, 1.9 trillion yen of which was ICs, and today it is responsible for one-third of world production. Thus, dispelling the myth that military demand is the driving force of high technology, Japan's semiconductor technology has reached the world's top rank. Although in the long run great developments are expected in terms of both technology and demand, the industry is now facing some problems which have to be overcome.

Japanese enterprises have achieved competence in producing standard goods such as memory, but when it comes to circuits for operation they are still behind the US. This design and development ability in the area of custom use (ASIC) is expected to be strengthened. In the field of design, attempts have been made to induce CAD since the latter part of the 1960s.

Second, the advance in integration – in other words, the development of microscopic processing technology – renders the existing manufacturing equipment obsolete in a shorter time. Equipment needs renewing every three years – once in five years is no longer sufficient. Semiconductor manufacturing, which has already become a capital-intensive industry, should be able to renew its equipment within a short time, and also carry out thorough automatization.

Third, in the shift from the LSI to the VLSI stage – in spite of the fact that the industry is undergoing rapid growth – there is an undulating discrepancy in the balance of supply and demand. Some call it the "silicon cycle," which undulates at intervals of about three to four years.

To enter the semiconductor business as a newcomer used to be thought difficult because of the need for technology accumulation and the high investment in equipment. However, in Japan, new assembly makers continue to join the business. On the other hand, there are examples of withdrawal. For this reason, more attention should be paid to the fact that in addition to technology, economic and managerial factors are influential in the semiconductor business.

Fourth, semiconductor goods have increased in importance in Japan–US trade. The recession which started in the latter part of 1984 increased friction between the two countries, and made Japan increase imports and overseas production. It is now necessary for the semiconductor business to be run with a worldwide perspective; Japan has to give due consideration to the situation in developing countries.

In the micro-electronics revolution, semiconductor goods should not be considered as only part of the electronics industry itself, but as something that concerns all industries, taking into account the general trend in electronization. Semiconductors should be regarded as the raw material for all industries, and not only as an end-product. The trends toward digitalization in electronics and the desire to make every machine intelligent are crucial factors. Semiconductors are important wherever information is made use of in society.

Japan's experience and Asian perspectives

Transfer of high technology to developing countries

The development of electronics industrial technology in the last three decades is a historical event comparable to the Industrial Revolution in the late eighteenth century. One major difference between the current electronics technological revolution and the Industrial Revolution can be found in the intensively rapid progress in product and process innovations occurring today.

The memory capacity of microchips, for example, has increased from 4K bit dynamic random access memory (DRAM) in the mid-1970s to 16M bit DRAM by the mid-1980s – a dramatic four-thousand-fold change within a decade! This electronic technology has been developed in industrial countries, particularly in the USA and Japan. As a result, the technology gap between industrial countries and the third world in general has become much wider in the 1980s than in the 1960s.

The rise of Japan as one of the major industrial countries, and that of the Republic of Korea as one of the newly industrialized countries (NICs), was made possible by the international transfer of technology. The industrialization of the world, we may argue, is indeed a historical process of international transfer of technology from more advanced to less developed countries. Therefore, the current electronics technological revolution could spread to the developing countries in the future. One could thus pose the legitimate question of how and when the international transfer of electronics technology, or, more generally, "high technology," will take place in the third world.

The possibility and desirability of transfer of high technology to developing countries will be discussed in terms of microchip technology. Microchips are the most essential inputs for the electronics technology; the product and process innovations in the microchip industry generate a wave of product and process innovations in the electronics industry as a whole, which in turn create another wave of innovations in related industries. In this regard, we can represent microchips as a high-technology industry. In the following sections we shall discuss, first, the basic characteristics of the industry and, second, the relevance of the microchip industry to developing countries. Finally, a tentative conclusion concerning the transfer of high technology to developing countries in the short run will be drawn.

Major characteristics of the microchip industry

Short product cycle

The intense competition for the development of new microchips among the leading electronics firms in the world was observed at the International Solid State Circuit Conference (ISSCC), which was held in New York in February 1987. International Business Machines Corp.

(IBM) announced that it has designed and produced a significantly more powerful computer memory chip that can store more than four million bits of data. This 4M bit DRAM chip, however, was not the monopoly of IBM. Other companies, including Texas Instruments Inc. and several Japanese concerns, also unveiled their successes in the development of a 4M bit DRAM chip. At the same meeting, Nippon Telegraph and Telephone Corp. (NTT) announced that it had already produced a 16M bit DRAM chip. This new memory chip, which is 8.9 mm by 16.6 mm in size, can integrate about 40 million components. Its cycle time is 180 nanoseconds (one nanosecond is one billionth of a second). This means that the chip requires only 0.4 second to read and write 64 pages of a newspaper.

In 1987, the 256K bit DRAM chip was the most widely used computing memory chip, and the mass production of a 1M bit DRAM chip by some Japanese and US electronics firms had just begun. The NTT's epoch-making breakthrough was evidence that the technological lead would soon be captured by other major electronics firms. Thus, the innovation of a new 4M bit DRAM chip, and even of a 16M bit DRAM chip, will accelerate the shift of the product cycle toward superpowerful memory chips.

It is said that a lifetime of a unit of productive facilities for certain types of microchip may be as short as four years owing to the very rapid product cycle. This machinery and equipment for microchip production will become obsolete within a short time-span, and chip manufacturing firms are forced to continue reinvestment for new production facilities in order to survive in the face of keen competition. LSI manufacturing facilities are in the order of 36–39 billion yen. Nevertheless, Japanese electronics firms are tending to intensify their investment in production facilities for microchips. The value of IC sales increased from 10.8 billion yen in 1975 to 179.8 billion yen in 1984, while IC-related investments rose from 1.1 billion yen in 1975 to 76.3 billion yen in 1985. The ratio of IC-related investments in the value of IC sales thus rose from 10.5 per cent in 1975 to 42.4 per cent in 1984. This enormous financial capacity to invest is an indispensable condition of survival in an electronics revolution of which the quick product cycle is one of the major characteristics.

Rapid process innovation

The second important characteristic of the microchip industry is found in the rapid development of process innovation. Thus, in the product innovation from 64K bit DRAM to 256K bit DRAII, and further to

327

1M bit DRAM, every processing step required a new technological breakthrough. Many process innovations take place along with the product cycle.

Inter-industrial technological linkages

Innovations in processing technology require considerable interactions with technologies developed in other industries outside electronics. The combination of technological developments in, for example, the construction, chemical, and textile industries makes it possible to fabricate microchips. It should be mentioned that related development in the technology of machine and electronics engineering created "mechatronics" engineering. The development of microcomputers was combined with machine engineering, and produced many kinds of NC machinery, machining centres, and other types of super high-precision instruments. The process innovations of the microchip industry would not have been possible without simultaneous technological development in mechatronics. In this regard, the microchip industry is the one which needs to apply technological progress in other industries. The inter-industrial technological linkages are of vital importance to product and process innovations in the microchip industry.

The extent of inter-industrial linkages can be shown by the input of other industries into the production of microchips. In 1984, microchip production amounted to 2,600 billion yen in Japan. A total of 400 firms in related industries played supporting roles in the industry, with annual sales of 1,000 billion yen.[11]

The growth of chip fabrication machinery and equipment in Japan has been very rapid, resulting in the quick import substitution of those production facilities. Without the available technological support by related industries, domestic manufacture of and innovations in microchip production facilities could not be attained.

Substantial expenditure on R&D backed by capable human resources

Technological progress in the microchip industry has been made possible in Japan by the massive investment in R&D which has been carried out jointly by the private sector and the government. Electronics firms in Japan spent nearly 215 million yen – 19.9 per cent of the sale of microchips – on R&D activities in 1975. Their expenditure on R&D steadily increased to 2,549 million yen, 17.5 per cent of the sale of microchips, in 1985. The number of R&D-related employees in microchip manufacturing firms increased from 2,337 in 1973 to 5,695 in 1984.[12] This figure does not include research staff in universities and public agencies. Therefore, the actual number of persons engaged in IC-related R&D in Japan is certainly much larger.

Many major electronics firms in Japan have established their own R&D institutions, and they have been very active in making technological innovations in order to compete with other firms in the industry. One thing to be mentioned here is that electronics firms are extremely competitive among themselves, but at the same time they tend to cooperate in the R&D for national projects. This may imply that major innovations are too expensive to be carried out by an individual firm.

Facing the serious technological challenges by Japan, US microchip manufacturers have also decided to form a billion-dollar cooperative manufacturing venture, the Sematech. This will be jointly funded by government and industry, and designed to restore the international competitiveness of US microchip manufacturers. This fact alone demonstrates what extraordinarily large R&D expenditures are needed to maintain a leading position in the era of the electronics revolution.[13]

International transfer of high technology to developing countries in East and South-East Asia

Japan's technological development, along with the electronics revolution, has again proved the usefulness of the strategy of borrowing technology from abroad. Japan imported electronics technology mainly from the US, and went through the process of adaptation and indigenization of imported technology. It is not surprising, therefore, to see that many Asian countries hope to learn lessons from Japan. They have already entered the phase of primary export substitution, and have steadily increased their exports of manufactured goods, including consumer durables and even electronic goods. With this level of industrial development, transfer of high technology from abroad is one of their major national concerns.

The Republic of Korea has already emerged as an exporter of VLSI, and Taiwan, Singapore, Malaysia, and Thailand are also exporting microchips. It appears that the Republic of Korea has now the capacity to become one of the world's major VLSI exporters. In this regard, it is interesting to examine the Korean strategy of electronics development. Table 7 shows the Korean perspective on technological progress toward the year 2000, and divides the industrial sector into three groups. Group 1 consists of electric appliances, iron and steel, petrochemicals, textiles and sundries. Group 2 comprises

Table 7. Korean perspective on technological development towards the year 2000

Product cycles in industrial countries	Korea's domestic technological development phases				
	Imported and indigenized technology	Improvement of indigenized technology	Early phase of imported technology	High technology	Future technology
Entry					Group 3 — Computers and communications, Biotechnology, Robotics
Growth			Group 2	LSI, Mechatronics, Aviation	
Maturity	Group 1	Automobiles, Shipbuilding	Electronics, High-precision machine tools, Advanced chemicals		
Decay	Electric appliances, Iron and steel, Petrochemicals, Textiles, Sundries				

Source: Korean Institute of Development Research, *A Long-term National Development toward National Development*, p. 122. Requoted from M. Saito, *International Political Economy of Transfer of Technology*, Tokyo: Toyo Keizai Shimpousha, 1986, p. 49.

automobiles, shipbuilding, electronics, high-precision machine tools, and advanced chemicals. Group 3 covers LSI-related electronics, mechatronics, aviation, computer and communications, biotechnology-related industry, and robotics.[14] The classification of industries into these three groups gives an insight into the Korean perception of technological progress. As an NIC, the Republic of Korea has shown remarkable progress in establishing a strong industrial base with the clear aim of catching up with technological developments in the industrial countries.

In the Republic of Korea's technological development, however, there appears to be a continuity from light consumer industry to heavy chemical industry, and further to high technology industry. On the other hand, the time-span in which Korea mastered the necessary technology for certain industries is astonishingly short. The process of learning-by-doing to acquire higher levels of technology was condensed into a short period, and she has been accelerating her national capability to absorb higher levels of technology.[15] What we should note here is that Korea did not try to absorb the electronics technology before she had mastered the technology of manufacturing automobiles and shipbuilding. Another point to be stressed is that Korean industrial development took place along with the formation of powerful company groups such as Hyundai, Samsung, and Daewoo. These company groups played decisive roles in financing and industrial investment and also in mobilizing entrepreneurial resources.

The ASEAN countries have also made notable developments in industrialization. They have entered into the phase of primary export substitution in which the export of manufactured goods has been steadily increasing. Their industrial progress has been facilitated by an international transfer of technology. Although the level of industrialization varies among them, it would be safe to argue that almost all the ASEAN countries have already developed industries which can be classified into the group 1 of the Korean case (table 7). However, it is hard to believe that they have firmly established the industrial base of the group 2 Korean industries. Nevertheless, most of the ASEAN countries have begun to develop high-technology-based industries. In some countries like Malaysia and Singapore, the international transfer of high technology seems to be a national project.

The transfer of high technology to these countries is technically feasible, as shown in the foreign direct investments undertaken by major electronics multinational firms in these countries. Either as subsidiaries of multinationals or as in joint ventures with local firms, the microchip industry was established in the ASEAN countries. In most cases, however, the plants were constructed on a turnkey base; the machinery and equipment were directly imported from parent multinationals; necessary technical staffs were dispatched from them; and the products were exported with their help.

More importantly, the microchip industry in the ASEAN countries is primarily engaged in the downstream operations of microchip fabrication. Most of the plants are assembly plants, which require relatively unsophisticated technology in the whole range of technological packages in the industry. In essence, these plants are used as instruments for the offshore operations of multinational electronics firms. It is therefore inevitable that the microchip industry tends to create a "technological enclave" in these countries.

In the previous section, we pointed out the four major characteristics in the microchip industry: (1) a short product cycle; (2) rapid process innovations; (3) inter-industrial technological linkages; and (4) substantial expenditure in R&D backed by capable human resources. At the moment, all the ASEAN countries appear to face considerable financial difficulties in developing the microchip industry. First, it is not at all easy to continue new investment, given the short product cycle of microchips. In fact, the cost of investment in a larger-capacity microchip has substantially increased. Furthermore, the unit price of microchip fabrication equipment rises along with the increasing integration of technological sophistication. The present level of available inter-industrial technological linkages in the ASEAN countries seems to be insufficient to allow them to embark upon the import substitution of the machinery and equipment for the fabrication of microchips. Thus, import dependency will continue, at least in the short run. The S&T infrastructure in terms of the availability of capable human resources and R&D institutions has still to be developed in these countries.

Conclusion

The international transfer of high technology to developing countries has to be examined country by country, since the level and structure of the industrial sector are, to a considerable extent, different among them. There are many means of transfer of high technology: (a) the import of high technology by licensing agreement; (b) the import of high-technology-related commodities, machinery and equipment, and plants; (c) the establishment of multinational subsidiaries; (d) joint ventures with multinational corporations; (e) subcontracting with multinational firms; and (f) the participation in international R&D projects.

Available options are limited by the recipient country's absorptive capacity for high technology. The Republic of Korea appears to have wider options than the ASEAN countries. She has already established the technological base for licensing production of microchips, whereas many countries in the ASEAN region still depend on foreign direct investment in the form of either subsidiary arrangements or joint ventures. Therefore, their immediate requirement is to strengthen their absorptive capacity for high technology rather than to embark upon the large-scale investment necessary to develop high-technology industries.

This view may conflict with the national self-image of the ASEAN countries. The Korean experience tends to support this tentative conclusion. Only after the Republic of Korea had mastered the necessary technology related to the group 1 industries did she venture to catch up with high technology.

Self-reliance in S&T for national development does not rule out the strategy of borrowing technology from abroad. Nevertheless, this strategy may end in failure if a country borrows technology which cannot be effectively utilized; this will only result in increasing dependency on foreign technology. A step-by-step approach in the international transfer of technology and a massive national effort to shorten the time requirement for mastering imported technology could be a realistic option for many developing countries, including those in the ASEAN region.

Japanese multinational enterprises and their role in technological self-reliance in Asia

Multinational enterprises (MNE) are generally considered to be the most important agents for the international transfer of technology, because they were and are the major creators of new technology, and also because they monopolize the main intellectual properties relating to industrial technology. The principal channels for transfer of technology used by these enterprises are licensing and direct investment. The developing countries, however, often lack the capability to use this manufacturing technology without the transfer of management skills associated with foreign investment. Japan is a rare case of a success story in which a country made effective use of foreign licensed technology without direct foreign investment.

The owner of the industrial technology may choose to invest with a wholly owned subsidiary or a controllable joint venture in which the technology provider holds more than 50 per cent of the voting stock. A choice of technology that is appropriate to receiving countries, the organization of management, the technology licensing fee, and even the sales area for products made by local plants are decided by the headquarters of MNEs. For the directors of any MNE, profit comes before the interests of the local community, technology transfer, and the self-reliance in technology of host countries.

The technology supplied by MNEs and the impact of transfer modes upon host countries' long-term technological growth and development have been criticized by them. United Nations organs such as UNCTAD (United Nations Conference on Trade and Development) have attacked the MNEs' attitude on technology transfer. UNCTAD has proposed an "International Code of Conduct on the Transfer of Technology" aimed at restricting the monopolistic power held by MNEs in the developing countries. The code of conduct aims at obtaining effective assistance for developing countries in the selection, acquisition, and effective use of technologies appropriate to their needs.

However, the focus of technology transfer problems is shifting from coping with the monopolization of technology by MNEs to the development of indigenous technology in developing countries. How to enhance local technology with the help of MNEs is an important question. In the following section, we will analyse the Japanese MNEs' behaviour and their impact on technological development in host countries.

Japan's direct foreign investment in Asia

The direct foreign investment activities of the Japanese enterprises can be divided into the following five stages.

The first period: 1951–1962
During the first half of the 1950s, Japan's foreign investment averaged annually less than $10 million, largely as a result of the limitations imposed by the precarious position of her own foreign exchange reserves. The amount of investment in Asian countries was meagre throughout the 1950s, in comparison with that in the Middle East and Latin Amer-

Table 8. Value of Japan's direct foreign investment approved by the Japanese government (US$ millions)

Fiscal year	Direct foreign investment	Cumulative value	Growth rate %
1951–61	447	447	
1962	98	545	21.8
1963	126	671	23.1
1964	119	790	17.7
1965	159	949	20.1
1966	227	1,176	23.9
1967	275	1,451	23.4
1968	557	2,008	38.4
1969	665	2,673	33.1
1970	904	3,577	33.8
1971	858	4,435	24.0
1972	2,338	6,773	52.7
1973	3,494	10,267	51.6
1974	2,396	12,663	23.3
1975	3,280	15,943	25.9
1976	3,462	19,405	21.7
1977	2,806	22,211	14.5
1978	4,598	26,809	20.7
1979	4,995	31,804	18.6
1980	4,693	36,497	14.8
1981	8,932	45,429	24.5
1982	7,703	53,132	17.0
1983	8,145	61,277	15.3
1984	10,155	71,432	16.6
1985	12,217	83,649	17.1

Source: Up to 1977: Terutomo Ozawa, *Multinationalism: Japanese Style*, Princeton, N.J.: Princeton University Press, 1979, p. 12. 1978–1985: MITI, *Kaigai toshi tokei souran: dai 2-kai kaigai jigyou katsudou kihon chousa*, Tokyo: MITI, 1986.

ica, because of the anti-Japan feeling that still prevailed in Asia. But the reparations paid by the Japanese government to some Asian countries offered Japanese private enterprises an opportunity to resume or to initiate open business activities in those countries (table 8).

The second period: 1963–1967
This period coincided with Japan's entry into OECD and the change of her status to an IMF Clause 8 country. During the early 1960s, Taiwan, Thailand, Hong Kong, and Singapore began to emerge as the host countries for Japan's offshore ventures in both commercial and

manufacturing operations. The change of development policies in most Asian countries from import substitution to an export-oriented policy offered favourable conditions to Japanese investors. A typical example was the establishment of the Kaohsiung Export Processing Zone in Taiwan in 1966. The amount of direct foreign investment exceeded the $200-million level by 1966. This increase was caused mainly by the growing investment in Asian countries, in particular Taiwan, Hong Kong, Thailand, and Singapore.

The third period: 1968–1971
This period saw the increasing liberalization of direct foreign investment on the basis of the accumulated trade surplus. The largest amount was invested in North America and the second place was now taken by Asia. Japan's normalization of diplomatic relations with the Republic of Korea took place in 1965. Korea opened the Masan Free Export Processing Zone in 1970, and offered very attractive conditions to foreign investors. Thus, Taiwan and the Republic of Korea came to occupy important places in Japanese enterprises' offshore production for export to other countries, especially in textiles and electricals.

The fourth period: 1972–1977
In June 1972, all foreign investment was completely liberalized and this, together with the appreciation of the yen, gave momentum to the phenomenal increase in the amount of investment. Even the Middle East war of October 1973 (Yom Kippur War) did not stop the increased investment flow and only slowed it down temporarily, as shown in table 8. The characteristics of Japan's direct foreign investment in this period were that the investment in advanced countries was mainly for commercial activities while that in Asian countries was mostly for manufacturing.

The fifth period: 1978 onward
The beginning of this period signifies the intensification of the "trade war between Japan and Europe" and "trade friction between Japan and the US." A sharp increase in the value of the yen relative to the dollar began in 1977, and in October 1978 reached the record rate of 176 yen per dollar. These two factors encouraged Japanese enterprises to invest in North America and Europe to circumvent protectionism.

In the 1980s, the pattern of Japanese foreign investment has changed noticeably. The share of investment in resource development and labour-intensive industries has been decreasing, and accordingly the relative importance of developing countries in Japan's total foreign investment has begun to diminish. On the other hand, Japan's investment in advanced countries has been substantially increased, especially in the field of technology-intensive industries such as electronics and transport machinery, in addition to her previous investment in non-manufacturing industries.

The decreasing share of Japan's investment in Asian countries has been caused by the following factors: (a) the problem of debt accumulation in some countries in the third world; (b) the completion of a round of investment in import-substituting industries; (c) the stagnation of the demand for primary products; (d) increasing political instability.

In addition to these, the MITI survey[16] lists the following factors, among others, as sources of the difficulties experienced by Japanese investors in Asia: (a) intensification of sales competition; (b) inflation; (c) difficulty in securing labour, both in terms of quantity and quality; (d) restrictions on the employment of foreigners and the obligation to employ local personnel; (e) restrictions on the import of raw materials and parts; (f) localization of equity capital demanded by the host country's government; and (g) restrictions on the mobilization of local capital.

Motivations for direct foreign investment

The MITI survey of Japan's direct foreign investment in 1978 gave information on what motivated Japanese enterprises to invest in foreign countries.

"To secure and expand the market" came first. Then followed: "to diversify and to internationalize business activity," "favourable labour conditions," and so on. The order of priority given to various motivations differed, of course, from region to region, and, even in the same region, from industry to industry. In the case of Asia, the factor of "favourable labour conditions" occupied a prominent position after "to secure and expand the market." This contrasted sharply with the case of the US, where the labour factor had almost no importance.

The labour factor played a more important role in light manufacturing industries such as textiles, electricals, etc., than in other industries. During the 1960s, the labour shortage problem in Japan forced a lot of medium and small-scale enterprises in textile and electrical industries to invest in Asian countries where the wage rate was much lower than in Japan. But this situation seems to have begun to change.

Table 9. Motivation for direct foreign investment in ASEAN (percentages)

Motivation	Time of decision	Present (1986)
To secure market	30	40
To localize production in the face of host country's import restrictions or ban	14	12
To secure production base as part of global business strategy	13	19
To expand business activities	11	12
To utilize cheap labour	10	7
To take advantage of host country's favourable investment promotion measures	9	3
To meet requests from partners	9	6
To secure raw materials	3	0.5
Others	1	0.5

Source: Japan Overseas Enterprises Association, *ASEAN shinshutsu nikkei kigyou ni okeru gijutsu iten: sono mondaiten kaimei to kaizen no tameno teigen*, 1986, pp. 46–47.

A survey of 79 enterprises by the Japan Overseas Enterprises Association on reasons for investing in ASEAN countries at the time the decision was taken, together with a present reassessment, is shown in table 9.

The table reveals that the importance of the host country's favourable investment promotion measures as a pull factor for foreign investment has been decreasing, apparently because of the increasing demands for localization put forward by the host country's government. The desire to secure and to expand the market was maintained throughout the period.

One significant change that should be noted is the importance attached to the motivation to secure the production base from the viewpoint of global business strategies. The emphasis on this factor seems to have been getting stronger, especially in the face of the slow expansion of the world market since the second oil shock. Even in Asia, this factor has gained more importance than before, as is clear from the data from another survey conducted by the Electronic Industries Association of Japan on the motivation of 74 electronic enterprises for investing in ASEAN countries. According to this survey, four major incentives, in descending order of importance, are "export to another country," "to secure host country's potential market," "to utilize cheap labour", and "host country's favourable policies for foreign investors."[17] One interpretation of this order of priorities could be that the aim is first to secure the host country's market by taking

338

advantage of its favourable government policies and cheap labour, and then to export to another country, using not a Japanese brand-name but one from the host country.

This arrangement seems to have been not wholly unwelcome to the host country, because the change of strategy on the part of the Japanese enterprises coincided with the change of development policies from import-substitution to export-oriented ones on the part of the host country's government. This change of policies on both sides has contributed to some extent to the improvement of the balance of payments position of the host country and to the raising of its technological standards, allowing it to compete in the world market.

How to fit into local conditions

Once the decision to invest in Asian countries has been taken, one of the greatest problems faced by investing enterprises is the selection of appropriate partners. According to a survey conducted by the Japan Institute of Labour in 1986,[18] only 28.9 per cent of 76 sample enterprises claimed that there was no problem in selecting partners. Sixty per cent have experienced some difficulty in choosing collaborators, and gave the following reasons: (1) lack of rational managerial qualities; (2) inflexibility of political regulations; and (3) late entry into Asian countries.

Recognition of the first factor has resulted in a preponderance of overseas Chinese businessmen as partners of the Japanese affiliates in most Asian countries and regions. The Japan Overseas Enterprises Association gives the following data on the participation of overseas Chinese in joint ventures with Japanese enterprises in ASEAN countries: 36 (57.1 per cent) of 63 sample ventures in Indonesia; 30 (52.6 per cent) of 57 in Thailand; 29 (46.8 per cent) of 62 in Malaysia; 26 (44.8 per cent) of 58 in Singapore; and 8 (33.3 per cent) of 24 in the Philippines.

This might have serious repercussions on Asian people's attitudes toward the Japanese economic advance into these countries. Thus, Ozawa remarks:

It was a rational choice for many Japanese manufacturers to choose as their partners in their new local ventures those Chinese merchants who used to be their import agents. In general, the Japanese found much closer cultural and motivational affinities in the Chinese than in the local people. However, there is a deep-seated resentment against the economic dominance of overseas Chinese in such Asian countries as Indonesia, Malaysia, and Thailand. The

339

Japanese, by joining economic forces with this particular local interest, unwittingly have made themselves targets of local resentment.[19]

In the face of the changes in economic policy by developing countries in favour of increasing localization, notably Malayanization in Malaysia, Japanese enterprises intending to establish joint ventures in Asian countries should, in selecting their prospective partners, take into consideration not only the economic calculations, but also the possible political and social implications.

Localization of capital

Concomitant to the selection of partners is the problem of what type of affiliation should be chosen for a joint venture. The number of Japanese affiliates in Asia, by share of equity participation, is shown in table 10. Minority participation below 50 per cent accounts for 42.4 per cent of the total number of affiliates. It should be pointed out that this ratio is quite high if it is compared with the US. Two explanations can be offered for this.

The first is that general trading companies have been playing a prominent role in establishing and operating Japanese affiliates in developing countries. Trading companies lack manufacturing skills and are satisfied with acting as mediators between Japanese manufacturers and local partners in the establishment of joint ventures. Kojima and Ozawa point out:

The trading companies' willingness – and for that matter, the willingness of Japanese interests generally – to remain as minority owners may be related in

Table 10. Number of Japanese affiliates by share of equity participation

Share (%)	Number of affiliates	% of total
Below 25	126	8.8
25–49	483	33.6
50	116	8.1
51–74	165	11.5
75–99	146	10.2
100	400	27.8
Total	1,436	100.0

Source: MITI, *Kaigai toshi tokei souran; dai 2-kai kaigai jigyou katsudou kihon chousa*, Tokyo: MITI, 1986, p. 136.

part to the comparatively small stakes involved in most of these manufacturing ventures. Moreover, trading companies are in a position to exercise a great deal of managerial control without holding majority ownership, since they provide such critical services as supplies of inputs and working capital and access to market.[20]

The number of manufacturing joint ventures in Asia in which trading companies have been involved in one way or another was 364 in 1980. This accounted for about one-third of the total number of manufacturing joint ventures in Asia in that year.

Another factor to explain the large proportion of minority ownership in Japan's joint ventures in Asia is the regulations imposed on capital participation by the host country's government. Take Thailand, for example. There is no written rule that foreign capital should be less than 49 per cent. There exist, of course, exceptional cases which the government decides from time to time. Because of this rather ad hoc localization policy by the Thai government, the so-called fading-out process has been accelerating among Japanese affiliates. The average share of Japanese capital declined from 73 per cent in 1963 to 47.8 per cent in 1974, and then to 46.5 per cent in 1981.[21] In the case of the Philippines, the limitation imposed on foreign capital participation is, as a rule, 40 per cent. In other Asian countries, a more or less similar policy of localization of capital has been followed in order to facilitate the growth of local enterprises as well as to avoid domination by foreign capital. But a too rigid application of localization policy and the demands of too hasty localization can sometimes cause unnecessary conflict between two partners. It is typical of this situation that one of the Japanese complaints about the performance of affiliates is the local partners' inability to mobilize enough capital.

Localization of staff

The localization demand does not stop at the control of capital. It covers personnel and the procurement of raw materials and parts as well.

Some countries in Asia impose restrictions on the number of foreign nationals working in joint ventures by the system of work permits. This is one of the factors which hamper the smooth transfer of technology, particularly by limiting the entrance of the necessary number of engineers and managing staff required at the initial stage of a joint venture.

To meet the demand for localization put forward by the host country's government, most of the Japanese affiliates have been implementing measures to replace Japanese nationals by local staff, step by step. Generally three stages of localization of staff can be discerned.

At the first stage – 5 to 6 years after the start of the operation – control over ordinary operations can be entrusted to local staff. The percentage of Japanese nationals employed declines from 4 per cent at the time of the start of operations to 1 per cent toward the end of this stage.

At the second stage – 11 to 12 years after the start of operations – the Japanese staff assist only on emergency cases and the rest of the operation is the responsibility of local staff. But the decline in the number of the Japanese nationals slows down. At the third stage, the local staff master the operational techniques to deal with all the possible situations in the production processes of standardized products. The Japanese staff now specialize in managerial activities. At this stage, the number of Japanese staff is stabilized, with 15–20 per cent of directors, 20 per cent of administration staff, and 60–70 per cent of production and technology staff.[22]

The average number of Japanese staff per affiliate in ASEAN countries in 1986 stood at 4.8 men; 1.6 in management, 2.5 in technology, and 0.9 in other divisions.[23] This contrasts sharply with the situation in the US, where the number of Japanese nationals stationed elsewhere is much higher, particularly in commerce-oriented activities (trading companies). So the problem is certainly not one of numbers, but of the degree of participation in the decision-making process, in terms not only of production processes but also of managing all business activities. This topic will be taken up in the next section.

The general pattern of relationships between parents and affiliates in Asian countries that emerges from these surveys is the following:
1. Matters relating to production and technology are dealt with following the instructions sent by the parent company.
2. Matters relating to personnel and labour relations are handled according to the host country's laws and customs.
3. Matters relating to marketing are entirely entrusted to the local staff.

The shares of local staff in the total number of directors, division chiefs, and section chiefs are 34.3, 52.4 and 77.9 per cent respectively. It is considered very difficult to localize the posts of persons in charge of presidentship, finance and accounts, and technology. The main obstruction is deemed to be the difficulty felt by local staff in communicating with the responsible personnel in the parent company. This language problem will appear again in the next section as one of the factors impeding the smooth transfer of technology.

Performance of Japanese affiliates in Asia

A survey conducted in 1985 by the Japanese Institute of Labour discloses the following results about the performance of Japanese affiliates in ASEAN countries: 23.3 per cent of the total affiliates surveyed replied that they had been enjoying a stable amount of profit; 52.5 per cent said they were making a profit, but this was unstable; and 23.3 per cent were unable as yet to realize a profit.

It is noticeable that the earlier the operation is started, the more stable the profits become. On average, companies began to realize a profit after $5\frac{1}{2}$ years of operation. For example, 44.7 per cent of affiliates that started operation after 1976 have recorded no profits as yet.

Similar results were obtained by the Japan Overseas Enterprises Association survey, which shows that 61 per cent of sample enterprises in ASEAN countries are making a profit, while 23.6 per cent report no profit.

Three main factors are responsible for this less than satisfactory performance.

The first is the high rate of defective products turned out in various production processes. If the level of Japanese enterprises is put at 100, the level of about half the affiliates ranges from 110 to 120. Twenty per cent of affiliates are said to be at the same level as Japan, and 16.6 per cent are assessed as better than Japanese enterprises.

The second factor is the low labour productivity. 76.7 per cent of affiliates surveyed are considered to be inferior to the Japanese standard in the matter of labour productivity. Of them, 40 per cent reported that their labour productivity was lower by 30 per cent than the Japanese standard.

The third element for the rather unsatisfactory performance of Asian affiliates is the compulsory local content requirement imposed by the host country government, of which mention has already been made. The demand for hasty localization in terms of procurement of raw materials and parts can affect the quality of products and/or increase the price and decrease competitiveness in the market. This will in fact be detrimental to the interests of the host country.

Assessment of the performance of Japanese affiliates

How do the Japanese enterprises view their business activities in Asian countries? In what respects do they think they are able to contribute to the interests of the host county? The MITI survey gave the following results:[24]

1. Creation of employment opportunities (42.8 per cent).
2. Improvement of the balance of payments position of the host country by expanding exports (11.8 per cent).
3. Encouragement of local industries (9.9 per cent).
4. Education and training of skilled labourers and managers (9.7 per cent).
5. Promotion of the diffusion of technology (especially hardware) (8.5 per cent).
6. Improvement of the balance of payments position of host country by import substitution (7.4 per cent).
7. Development of natural resources (0.6 per cent).

This result is quite understandable if we recall that many of the enterprises which ventured into Asian countries are labour-intensive industries, such as textiles, electricals, etc., and that a fair number of them have taken advantage of favourable measures offered by the host country, like the right to operate in free export-processing zones.

It should be emphasized that matters relating to technology transfer (nos. 3, 4, 5, and 7 above) are recognized to be important contributions towards the interests of the host country.

Unfortunately, there is no detailed study on how local partners view their affiliation with Japanese enterprises. Though the number of samples is small, one such attempt has been made in the Institute of Developing Economies study on "Trade and Technology Transfer Frictions between Japan and Developing Countries." This study shows to what degree the partners of joint ventures are satisfied with the performance of their ventures.

From table 10, we can infer that the local partners feel more satisfaction with their joint ventures than the Japanese do. Evidence for satisfaction with the fairly successful transfer of technology (viewed from the point of the local partners) is the increasing international competitive power of such industries as textiles, electrical appliances, iron and steel, shipbuilding, etc., in some of the Asian countries and regions, especially the Asian NICs. In the case of Taiwan, out of 20 enterprises selected as ranking high in terms of export earnings, 13 are electrical appliances and electronics companies, and all 13 are either foreign subsidiaries or joint ventures. The purely local enterprises are prominent only in the light export industries such as textiles and apparel. Similarly, in the electronics industry in the Republic of Korea, the foreign subsidiaries and joint ventures, which account for only 16 per cent of the total number of enterprises, share 36 per cent of production and over 65 per cent of total exports.[25] The same tendency can be observed in the Hong Kong and Singaporean electrical and electronics industry, where the share of the US enterprises is bigger. Other evidence for the fairly successful transfer of technology is the rapid rise of the level of technology of the local industries, effected by the diffusion of technology from foreign enterprises and their affiliates. For example, in the electrical and electronics industry in Asian NICs, many of the local enterprises have made progress from simple production based on assembly to more technology-intensive fields such as VTR, computer software, and so on.

Similar results were obtained by the Japan Overseas Enterprises Association survey of 79 joint ventures. Twenty-two per cent reported that they were satisfied with their ventures, and only 9 per cent that they were dissatisfied. The rest replied that the performance of their joint ventures was neither good nor bad. It should be noticed that there is a discrepancy in the degree of satisfaction with joint ventures felt by the Japanese side and the local partners.

MNEs need more development of regional technology

In making direct investments abroad, MNEs have expressed little interest in developing "appropriate technology" or in supporting "indigenous technology development" when the needs of the business at hand have called for a significant departure.

Japanese MNEs are no exception. As mentioned above, Japanese MNEs prefer to decide technical matters in the parent companies, and are reluctant to allow localization of technological staff. However, they seem to have a different attitude to technology transfer and regional technology development. Three basic factors appear to adumbrate the future trend in the business strategy of MNEs.

The first factor is a licensing strategy. As Japanese MNEs move abroad, particularly to developing countries, they do not usually charge high royalty fees to partners. Sometimes they sell their know-how without fees. Because Japan is still a developing country in terms of technological development, it has only a limited amount of technology that it can charge for. Japan still has a big deficit in its technology trade balance.

The second factor is that the Japanese MNEs' superiority lies in process innovation. Because Japanese technology development is especially weak in basic technology, Japanese MNEs try to concentrate their efforts in local production by the transfer of process innovation.

Japanese manufacturing industries have earned a good reputation for the quality and reliability of their products. It has been shown that high productivity with high quality has enabled Japanese automobiles and electronics products to dominate the world market. Process innovation, such as the quality circles (QC) movements, has contributed to this success. The focus of Japanese MNEs' technology has shifted from new product development to process optimization and cost reduction.

In the American MNEs, R&D efforts centre on the exploration and development of new technologies and products. In Japanese MNEs, in contrast, the emphasis is on finding new ways to make products economical.

Process innovation, such as QC circle activities, needs more cooperation with workers and management and collaboration with local industries. These technological developments make it possible to use workers' suggestions and proposals from the shopfloor.

The third factor important to the regional development of technology is the intimate relationship between parts industries and subcontractors.

Japanese competitiveness in the world market is strong in specific manufacturing industries such as automobiles, TVs, cameras, and other electronics. These assembly-makers generally rely heavily on parts makers and subcontractors. MNEs that produce TV sets and automobiles buy and assemble their parts from local producers. To maintain high quality and high productivity in the final product, they have to utilize the same quality-control systems and highly organized delivery systems with subcontractors. MNEs that operate local plants need to develop more regional technology and help the parts makers by the rapid dissemination of technology.

Japanese MNEs have changed and are changing their policies towards better cooperation with local entities in terms of a global business strategy. However, at the same time, they are aware of the difficulties of transferring technology to Asian countries.

A Japan Overseas Enterprises Association survey provides a fairly comprehensive list of factors which are considered to obstruct the smooth transfer of technology, from both the Japanese side and the local side.

The factors from the Japanese side, in descending order of importance, are as follows:

1. Low level of technology in general.
2. Shortage of managerial staff.
3. Low quality and high price of parts locally procured.
4. Insufficient recognition of the importance of quality control.
5. Insufficient qualification of technicians and engineers.
6. An unwillingness to share the acquired knowledge of technology with others.
7. Insufficient recognition of the importance of maintenance of equipment and machinery.
8. Low quality and high price of equipment and machinery locally procured.
9. Low quality and high price of raw materials locally procured.
10. Insufficient financial capacity of partners.
11. High turnover rate of local employees, especially engineers and technicians who have been to Japan for training.

The factors from the local side, again in descending order of importance, are:

1. Lack of communication due to insufficient knowledge of languages, both English and local.
2. Slow speed of technology transfer (e.g. localization).
3. Shortage of appropriate textbooks and manuals.
4. Unreasonably high cost of technology.
5. Inadequate system of education and training.
6. Too narrow specialization of the Japanese staff sent from parent company.
7. Period of stay of the Japanese staff dispatched from parent company too short.
8. No provision of latest technology.
9. Limitation of sales area.
10. Lack of leadership among the Japanese staff sent from parent company.
11. Insufficient effort made by the Japanese staff.
12. Calculating nature of the Japanese.
13. Technology transfer often accompanied by supply of high-priced equipment and parts.
14. Slow decision-making on the Japanese side.

If we compare carefully the complaints put forward by both parties about the seemingly unsatisfactory state of technology transfer, we notice one important difference of opinion. It is clear from the above list that almost all the complaints made by the Japanese are related mainly to the local culture in general or to the strict government regulations on localization of production which have direct implications for individual enterprises. As against this, the complaints put forward by the local partners are more specific, related to the actual process of technology transfer, and many could be solved at the level of ordinary operations if due consideration were given to them.

This discrepancy in the perception of the state of technology transfer is apparently caused by differing views on the concept of self-reliance in technology. This means that the complaints of both sides are not in the nature of a contradiction or confrontation, but in the nature of a misunderstanding, or rather an inadequate mutual understanding. To resolve this unhappy state of affairs, it is necessary for both sides to share a common perception about self-reliance in technology.

If self-reliance in technology in manufacturing industry means the attainment of the ability to produce standardized products designed by the parent enterprises, it can be asserted that many Asian countries have reached or nearly reached this stage in some fields of industry. But if we define self-reliance as the attainment of the ability to modify and improve standardized products to suit local conditions and, further, to design new products adaptable not only to the host country's conditions, but also to the conditions of the country for which exports are destined, then it becomes clear that only a few countries in Asia have achieved this. For this to be achieved, a huge amount of R&D expenditure is required, targeted at directly related production processes as well as at many ancillary fields and processes.

The problem does not end here. Localization of manpower (symbolic manifestation of the completion of technology transfer) has two aspects: one is related to production processes, and the other to management. Staffing the production processes with local technicians and engineers might be easier than entrusting management to local managers. The management know-how and information accumulated by several generations of able and forward-looking businessmen is indispensable to the nurturing and maintenance of an efficient cadre of managers who are capable of coping with the ever-changing world of business and of operating firms efficiently and profitably. If this aspect is forgotten, progress will stop at the stage of imitation or a little beyond, and will never reach the stage of creation.

Technological self-reliance in Asia: In search of a new international technology order

Some aspects of self-reliance in technology

Among third-world nations, a new international technology order has been awaited as a key factor in securing economic self-reliance. A new international technology order would build an international horizontal network of collaboration between equal partners, in contradistinction to the current vertical technological order.

However, this kind of structural change is still in the future as far as developing nations are concerned. They are deeply dependent on foreign technologies for their development, and self-reliance in technology is still far away. Self-reliance does not mean autarky in technology but, on the contrary, a firm basis for the absorption of any new technology useful for national development; this would be the first step towards a new international relationship and structure.

There are sharply different opinions over the means and methods to achieve this. In our view, the way forward is to develop a dialogue for Asian development and to invite open criticism in order to deepen mutual understanding and resolve problems.

Self-reliance as the basis for a new international technological order

Many development economists draw the attention of policy makers to three fundamentals: (1) domestic capital formation; (2) resources development and machine production at home; and (3) a resourceful army of manpower. In addition to this, we must recognize a fourth – the accumulation and open use of information. One of the problems for development in the third world is the unreliable supply and/or delivery of source materials and component parts, which in addition to being late are often both short in quantity and poor in quality. Punctuality and precision of the linked sectors and services are all indispensable for industrial development, not only for national development, but also for the international network of development. For instance, the Indian standard of technology in many sectors is extremely high, but their linkage remains at a very low level of development. A vicious cycle exists between mining, railways and power stations. Improvement in this fragmented linkage could be realized by an open exchange and a control system utilizing detailed and real-time information.

This kind of informational difficulty is seen in all societies during the stage of transformation from an agrarian to an industrial society. An efficient supply of detailed and precise information allows shop-floor engineers and business managers to take countermeasures and to guarantee their safe operation. This involves the training of local professionals, engineers, and managers, and these resourceful professionals must be recruited from locals who have internalized the country's culture and language. Of course, foreign professionals are helpful initially, but the final stage of naturalization should be left in the hands of locals. Only professionals with a firm grasp of their national fundamentals will have the flexibility to undertake the unprecedented mutual collaboration necessary for the new international technology network.

In this respect, it is encouraging to see recent trends in the industrialization of member countries. But it is still necessary to analyse the multinationals' strategies for technology transfer, as their essential aims are neither the self-reliance of the nation they are entering nor the new international technological order that we have discussed. Of course, there is no need to deny their favourable effect on the creation of job opportunities and on manpower development. However, in our view, it is simply not enough to build technological enclaves with no regard for the formation of a national network.

Japan's role in technological self-reliance in Asia

Today, Japan ranks second in the world, after the US, in the amount of economic assistance rendered and is the biggest holder of international financial assests. She is not only an economic power, but a technologically advanced country competing with the US in the area of high technology. Japan has, and will have, in the near future, a considerable capability to assist the economic development of LDCs and will also hold responsibility as a leader in the international economic order.

In Asia, Japan plays an important role in economic and technological development. However, as the Japanese role increases, we need to concentrate more on economic and technological cooperation with Asian countries.

On the basis of the two case-studies in which we described Japanese technological development and her efforts in technological self-reliance, we would like to make the following recommendations to Asian countries and to Japan in its cooperative efforts.

The first concerns the effective application of "Japan's experiences" of technological self-reliance. This does not mean that Asian countries should try to reproduce exactly Japan's experience, but that they should acquire the Japanese attitude towards new technology. Starting from imitation, Japanese industries have moved to creative imitation and on to self-reliance. Japan has to spend more time and money in teaching Asia not only technology itself, but also "Japan's experiences."

The second recommendation concerns the building of a technological infrastructure in Asia. This term, however, covers a wide range of topics, from education and training systems to information systems. In order to set up these infrastructures, the Japanese should concentrate their economic aid more in these areas.

The third recommendation relates to the role of Japanese MNEs in the technological self-reliance of host countries. We have described the Japanese MNEs' pattern of direct investment in Asian countries and their behaviour in technology transfer. Compared to other MNEs, Japanese companies tend to support efforts in regional technology development by a better use of local parts, industries, and subcontractors.

We are now in a position better to appreciate the impact of direct investment by MNEs, and should urge Japanese companies to promote technological self-reliance in host countries, thus keeping to the spirit of the international technological order.

The fourth recommendation concerns technological cooperation aimed at the technological self-reliance of Asian nations. Japan's policies for technological development were devised in her own interests. Every big high-tech development project is embarked upon for the sake of Japan. Recent joint projects with other nations in such fields as nuclear fusion for electric power generation and more efficient aircraft engines have been initiated only by highly industrialized, advanced nations. It is thus highly desirable that Japanese technical policy should emphasize more joint projects with developing nations in keeping with the ideal of self-reliance in technology.

Notes

1. Institute of Investment Economics, Development Bank of Japan, *Henbo suru kenkyuu kaihatsu toushi to setsubi toushi* (Changing Pattern of R&D and Investment), vol. 5-I, July 1984.
2. Economic Planning Agency, *Current Economy of Japan*, Tokyo: EPA, 1985, pp. 125–134.

3. Toshio Shishido, "Japanese Industrial Development and Policies for Science and Technology," *Science* 219 (1983) no. 21: 261.
4. Agency for Science and Technology, *Kagaku gijutsu hakusho* (White Paper on Science and Technology), Tokyo: Government Printing Office, 1969.
5. MITI, *Ko-shu gijutsu-dan chousa houkoku* (Reports on Imported Technology), Tokyo: MITI, 1965.
6. See note 5 above.
7. See table 2.
8. Toshio Shishido, "Japanese Technological Development," in T. Shishido and R. Sato, eds., *The Economic Policy and New Development Perspectives*, London: Auburn House, 1985, p. 208.
9. Ronald P. Dore, *Education in Tokugawa Japan*, London: Routledge & Kegan Paul, 1965.
10. Edwin Mansfield, "Technology and Productivity in the United States," in M. Feldstein, ed., *The American Economy in Transition*, Chicago, Ill.: University of Chicago Press, 1980, p. 563.
11. Nihon Denshi Kougyoukai, *IC shuuseki-kairo gaidobukku*, Tokyo, 1986, p. 53.
12. Nihon Denshi Kougyoukai (note 11 above), p. 54.
13. *Japan Times*, 6 March 1987.
14. M. Saito, *Gijutu iten no kokusai seiji keizaigaku*, Tokyo: Toyo Keizai Shimpousha, 1986, pp. 48–49.
15. T. Watanabe, *Seichou no Ajia, teiti no Ajia*, Tokyo: Toyo Keizai Shimpousha, 1985, pp. 56–61 and 117–121.
16. MITI, *Kaigai toshi tokei souran: dai 2-kai kaigai jigyou katsudou kihon chousa*, Tokyo: MITI, 1986.
17. Electronic Industries Association of Japan, *Kaigai seisan houjin list*, 1986.
18. Japan Institute of Labour, *Nikkei kigyou no gen-chika to roudou mondai: ASEAN chousa dai 2-ji houkoku*, Tokyo: JIL, 1986, pp. 18–19.
19. T. Ozawa, *Multinationalism: Japanese Style*, Princeton, N.J.: Princeton University Press, 1979; K. Ikemoto et al., *Nihon kigyou no takokuseki-teki tenkai: kaigai chokusetsu toushi no shinten*, Tokyo, 1981.
20. K. Kojima and T. Ozawa, *Japan's General Trading Companies: Merchants of Economic Development*, Paris: OECD, 1984.
21. Japan Institute of Labour, *Nikkei kigyou no gen'chika to roudou mondai: ASEAN chousa dai 1-ji houkoku*, Tokyo: JIL, 1986, p. 113.
22. JIL (note 21 above), pp. 36–39.
23. Japan Overseas Enterprises Association, *ASEAN shinshutsu nikkei kigyou ni okeru gijutsu iten*, 1986.
24. See note 16 above.
25. MITI, *Annual Report for International Trade* (Boeki hakusho), Tokyo: MITI, 1986.

7

The lessons from Asia: From past experience to the future

The countries of Asia vary widely in their social and economic characteristics, their recent histories, the different technology policies they have pursued, the different periods during which they embarked on industrialization, and the technology spectrum that was then available. The country studies that have been presented reveal the outcome of the interplay of these factors within the individual countries. Before attempting any generalizations about technological self-reliance and acquisition in Asia, it is useful briefly to recall the salient technology-acquisition experiences of the individual countries.

China

The Chinese communists, during the years of their guerrilla struggle,[1] had managed to achieve some manufacturing output. By the time they took power, therefore, they had by force of circumstances developed an implicit industrial policy. Yet in the years immediately following the revolution, they followed the Soviet model, although modifying it as a result of their own unique experiences. The Soviet formal system of science and technology, with its very rigid structure, was thus transplanted to China. Soviet technology was also transferred, largely through complete sets of equipment from the Soviet Union and Eastern Europe. After the break with the Soviet Union in the 1960s, the small number of imports of technology that then occurred came from Japan and Western European countries.

In the late 1950s, China attempted, through its "Great Leap Forward,"[2] a mass mobilization in technology acquisition, to catch up with some of the advanced countries. This partially involved "back-yard" technology, and was reminiscent of the period of guerrilla struggle. During the period following the break with the Soviet Union, a policy of "Walking on Two Legs," amounting to near-autarky, was strongly emphasized. Although this latter emphasis was later modified after the 1970s, the basic strategy of combining traditional labour-intensive methods with modern technology remained a part of general Chinese technological strategy.

After 1978, with the opening to the outside world, a major transfer of technology from abroad was attempted.[3] But the earlier, relatively rigid, Soviet-inspired formal system of S&T, which lacked active and organic linkages with the economy, still persisted. The R&D system lacked horizontal linkages with the economy, and the organization of the R&D institutes was over-centralized; consequently they could not develop their full potential. The system emphasized a technology "push" in the economy and was not generally responsive to demand. By the 1980s, the Chinese authorities were admitting that they were not only behind the developed countries, but also behind some of the NIEs.

The S&T system has consequently been considerably reformed in recent years, the new reforms aiming at developing organic linkages between the S&T system and the economy. These reforms have varied from changes in organization to personnel development.

By the early 1990s, at a time when new technologies such as information technology and biotechnology were beginning to transform the available technology spectrum in developed countries, China had gone through several learning phases in technology acquisition and had built up a considerable technology infrastructure. It was allocating roughly 1 per cent of its GDP to R&D, which, though low by developed-country standards, was high by developing-country standards, and by 1985 it had nearly 800,000 scientific and technical personnel.

India

In the case of India, a Soviet-inspired planning model was imposed on a mixed economy, the "commanding heights" of the economy and key industries being largely in state hands. This implied that, unlike in the Soviet Union, market signals as well as administrative guidance mediated the economy, giving rise to a potentially more flexible system.

Particularly since her second Five-Year Plan, India has emphasized the development of an S&T capacity. By the third Plan, an extensive

354

network of institutions for pure and applied research had been established, and the state was allocating about 1 per cent of total public outlays to the S&T programme. The different plans in India resulted in a sequence of somewhat differing emphases. These included: "self-reliance," primarily by building up the capital goods and machine-building sector, and the use of scientific research for the "benefit of the people" (the third Plan); maximizing the use of indigenous resources, employment, modernization, capacity utilization, and energy efficiency (the sixth Plan); and consolidating and modernizing the S&T infrastructure already built up and promoting certain front-line areas (seventh Plan, 1985–1990). The later plans recognized the lack of linkages between the S&T system that had emerged and the economy, and emphasized the need for this linking.

By the early 1980s, the Indian strategy had brought down the share of imports in gross capital formation to virtually a tenth of what it had been in the 1960s. This trend towards self-sufficiency was particularly noticeable in areas where the rate of technological change was not very fast. Indian firms had also developed a capacity for adapting imported technologies to the local context. And, although India still had a high percentage of illiterates among her population, she had expanded her higher and technical education to reach an estimated total of around 800,000 personnel in 1990.

However, although the relative achievements of Indian S&T were considerable, it was less successful that it could have been. The use of locally generated technology in large modern industries tended to be small and the national laboratories that had been developed had not become significant sources of industrial technology. Furthermore, despite financial incentives, the number of in-house R&D units was still small. There was no significant spill-over of S&T into the industrial economy. And, as in China, S&T tended not to be demand-driven, but to exist in isolation. In addition, studies have indicated that Indian scientific output has not been commensurate in quality and creativity with the numbers engaged in science.[4]

The Indian policy in S&T, like that of the Chinese, had been primarily one of import substitution. After debate, however, a general trend towards liberalization began in 1979 (as it did in China at around roughly the same time). The import of capital goods was liberalized. A significant increase in imported technology resulted after the 1980s, although the technology import policy still remained selective, targeting specific areas. New measures announced in mid-1991 pushed these tendencies further.

Republic of Korea

The Republic of Korea's independent industrial development has occurred primarily since 1945, in three stages. These were: a period of instability from 1945 to 1953, taking in the Korean War; a reconstruction phase from 1954, with an emphasis on import substitution; and a period of accelerated growth since the 1960s, corresponding largely to a switch to an export-oriented strategy. The Korean economy has been developed through the implementation of a set of plans.

In the initial post-Second World War period, especially during the Korean War, there was an inflow of technology from the United States and the United Nations. Because of the Korean War, and universal military service, the military became an important channel of technology transfer, which exposed the population to a variety of modern machinery.

As the Korean economic emphasis shifted, so did its technology acquisition programme. In the 1960s, it centred on the acquisition of import-substitution technologies for consumer goods industries and export-oriented light industries. In the 1960s a primary emphasis was given to the import of advanced technologies for the industries then being set up. A law to attract technology through foreign capital facilitated this, after some initial difficulties.

With the growth of the economy, the Korean Institute of Science and Technology (KIST) was established in 1966 to give technological assistance to the heavy, chemical, and other export industries. This helped to digest and absorb imported technologies and also to bring back Korean scientists working abroad. In the 1970s, many specialized industrial research institutes were established, largely spin-offs from KIST. The adaptation and improvement of imported technology was a primary aim of S&T policy during the third Five-Year Plan (1972–1976) and the fourth Five-Year Plan (1977–1981).

By the 1980s, the Republic of Korea had developed her industrial infrastructure to a considerable extent and was now competing with developed countries in some high-technology industries. Unlike, say, India or China, Korea did not have a strong ideological emphasis on S&T, but its S&T system had grown up pragmatically and organically in keeping with the requirements of the economy. Over a few decades the country had built up a viable technological acquisition programme that in the 1990s was beginning to compete in some cutting-edge technologies.

Thailand

The Thai report does not cover the major historical forces and constraints and the social environment discussed in the other reports. It briefly mentions some of the external influences that affected Thailand, such as the introduction of Buddhism and, later, aspects of building technology. The study mentions the beginning of the transfer of European-derived technology in the reign of Rama V. These first transfers occurred without Thailand undergoing direct subjugation by outside powers and largely on Thai initiatives, although these very initiatives were subject to a geopolitical environment in which outside powers were jostling for power around Thailand's borders.

The study concentrates more on the effects of the Thai national plans from 1961 to 1986. It notes, for example, that, under the first National Plan, "tremendous" change occurred in transportation, roads, and railways, that is, in the growth of Thailand's infrastructure. The number of educated people also increased significantly. During subsequent plans agriculture was also developed, more by extensive cultivation than by intensive effort, unlike in many other Asian countries during this period.

During the Plan periods, the private sector was encouraged to take part in industrialization under the close guidance and control of the government. But the import of technology that this entailed was done with only a minimum of mechanisms to select, control, and adapt the imported technology to Thailand's national needs.

Thailand's subsequent growth under rapid industrialization also gave rise to a strong rural-urban imbalance. The study notes the continuous growth of Thai industry during the period, but regrets that the industrialization was dependent on a heavy import component. The conclusion is that self-reliance was weak principally because of four factors: (1) science and technology was pushed from the top downwards, and there were (2) a lack of selection mechanisms to filter foreign imports, (3) inadequate efforts to evaluate R&D programmes, and (4) an insufficient link with the other sectors of the economy.

Philippines

If China, India, and Japan had explicit debates on science and technology, as well as on the relationships to the external world which were

357

translated in varying degrees to action, the Philippines case showed a lack of both significant debate and significant industrialization.

The Philippines report takes into account the external environment as well as the historical sequence of technologies. Self-reliance at the macro level for the Philippines is defined as the replicative capacity of "second-wave" technologies. The latter correspond to the technologies that were developed from the beginning of the Industrial Revolution up to the Second World War. Third-wave technologies comprise those that have been developed subsequently, such as information technology and biotechnology.

The report emphasizes that there was a large gap between the rhetoric of policy makers and the Philippines reality. A viable S&T policy was not a major concern of policy makers, and no serious attempt was made to introduce industrial technology. Moreover, a scientific community in the modern sense did not come into being as a functioning entity.

Cultural imperialism strongly influenced the nascent Philippines scientific community, the latter often looking over their shoulders at their mentors abroad. Furthermore, the interests of scientists were primarily in agriculture and medicine, whilst industrial research and basic research in the physical sciences were downplayed. The scientist as technician and taxonomist, rather than as discoverer, is an image that has persisted to the present.

The three major policy episodes in science and technology from the 1960s to the present had a tangible result only in education. The announced policy thrusts during these periods varied from import substitution in the 1960s, to the mission-oriented policies of the 1970s, and to the "demand-pull" strategy of the 1980s, which was accompanied by a Science and Technology Plan. Although some of these different attempts had some success, there was no significant departure from the basic framework of a dependent S&T system.

Thus, although the manufacturing sector in the Philippines grew in the initial period, it was built up on indiscriminate import substitution. The latter did not help to build organic linkages between industry and the rest of the economy. Furthermore, the proportion of scientists and engineers increased only marginally. Funding for R&D as a percentage of GNP also remained roughly stagnant.

The study attributes the weakness of the Philippines S&T capacity basically to an inability to break the colonial mould. Either the rules and regulations to filter the inflow of technology had loopholes, or the local bodies expected to do the screening did not have sufficient expertise. The Philippines is thus in contrast to China, India, Japan, and the Republic of Korea, where there was a national will to make a breakthrough in a technology-dependent world.

Japan

Of the countries studied in this research, Japan is the one that in a period of 100 years has made a successful transformation to the most advanced technological status. The parallels and contrasts with the other countries are therefore very instructive. Japan consciously opened up to the external world after the Meiji Restoration, and this step was taken only after considerable internal debate and a conscious awareness of, and control over, the process of opening up.[5] Thereafter Japan absorbed science and technology with the same zeal that she had shown in earlier centuries in imbibing mainland Asian influences such as Confucianism and Buddhism.

In this process, she passed through four stages: pure imitation (from the mid-1800s to the end of the nineteenth century); higher industrialization, adapting technology to local conditions (from the beginning of the twentieth century to the end of the Second World War); catching up with advanced technology (from the Second World War to the early 1970s); and from "imitation to creation" (from the early 1970s to the present).

In the first period, Japan was dependent solely on foreign personnel for S&T guidance. The teachers in higher S&T institutions were almost all foreigners, who were replaced gradually by locals.

At a later period, the government brought academics and businessmen together to form research councils in various fields. And this industry–academic partnership helped actively in the mastery of science and technology.

Again, during the third period of catching up with advanced technology, the importation of foreign technology was strongly encouraged, as importation reduced the commercial risks and uncertainties of newly created technologies. This allowed Japan to make rapid advances in the newer technology. These imports were under strong governmental guidance with regard to their effects on the international balance of payments as well as their technology composition.

In the fourth and final stage, that of creative technology, a high level of R&D expenditure has put Japan on a par with the US.

According to the study, the reasons for Japan's capacity for technology mastery include its high educational standard at the beginning of industrialization, its entrepreneurial spirit, its willingness to learn new technologies and to abandon old ones, the dual structure of Japanese industry, and the guidance given to small companies by large ones. In addition, other contextual factors led to Japanese self-reliance, such as the country's high savings and investment level, and a nationalism which initially did not encourage foreign capital, though money was temporarily borrowed from abroad.

The Japanese example reveals the success of a carefully nurtured pragmatic technology policy, introduced to meet the country's own felt needs, as perceived through its own culture. In fact, the Japanese study team recommends for other developing countries the use of re-sourceful professionals who have been recruited from locals and are fully imbued with the country's culture. The team emphasizes that only professionals firmly rooted in these national fundamentals can make the required flexible responses in the acquisition of foreign technology.

The geopolitical environment and the local socio-economic situation

What then, are the "lessons" that we can learn from these studies of different countries? The countries have varied in terms of size, political regime, stage of development, cultural and historical background, and socio-economic conditions. Important issues that are mentioned explicitly or implicitly in the country reports will be discussed below. Issues raised only in one or a few case-studies, but considered vital for the region in general, will also be elaborated and generalized. In addition, other broader aspects and contextual issues vital for self-reliance – particularly those relating to the newer technologies which have only recently come to the fore and are not covered by the individual studies – will be taken up. In the early 1990s the cumulative results of an epoch are being experienced and a possible geopolitical turning-point in technology witnessed. Hence the need for the larger picture.

A first "lesson" is that, in the more successful cases, the groups that led industrialization and technology mastery had a strong historical awareness of, and identification with, their countries prior to embarking on industrialization. They were also aware of the socio-economic environment in the world outside, especially in the then developed countries. With that background, they could help bridge the gap between themselves and the industrialized world. There was both a committed leadership and important political and cultural debates within the country on the need for, as well as the direction of, the path of industrialization.

To the category of groups that successfully led industrialization and the mastery of technology belong the original Meiji élite and its successors, including the post-Second World War Ministry of Trade and Industry (MITI) in Japan. In the case of China and India, those who debated and responded to the Western political and economic encroachment also belong to this category. Similarly, in the Republic of Korea the Japanese subjugation and the Korean War prepared the country for a major industrialization effort. All these four countries have succeeded to varying degrees in their industrialization efforts.

The Thai report does not mention the Thai debates, but the fact that it was not colonized and that its leadership could weigh options relatively freely were undoubtedly important factors in Thailand making not inconsiderable economic and industrial progress.

It is in the Philippines case that the effects of cultural colonization and the lack of argued-out positions on industrialization and self-reliance became apparent, resulting in the country's weak commitment to effective industrialization.

Formal S&T structure and industry

When a decision-making élite embarks on an S&T policy, it can choose from a range of perspectives. In the case of China and India, the formal structures of science developed as the outcome of a theoretical and formal analysis of the role of industrialization in development. However, the links between these S&T structures, which included academic and industrial institutions, remained weak and partly tangential. In recent years, both China and India have attempted to supply the missing links between industry and the formal S&T structures.

In the case of Japan and the Republic of Korea, the S&T structures and industry grew hand in hand, largely in a pragmatic fashion. Quite early on in Japan, there existed joint bodies of industrialists and scientists. The organic linkages between industry and agriculture meant that advancement in one fed the other. The Japanese emphasized applied research, that is, research closely linked to industry. Today the bulk of Japanese R&D takes place within the firms themselves rather than in the public sphere,[6] so that there is an immediate outlet for useful innovations, whilst, conversely, industry's demands are directly transferred to R&D groups.

361

In the Republic of Korea, the initial manpower training in the S&T sector aimed simply to provide technicians and engineers to operate and maintain industries. As industries developed, the S&T infrastructure developed, with constant interactions between the two. The Korean success in this strategy has been such that, in certain frontier areas such as chip manufacture, the country in the early 1990s is only a few years behind Japan.

In the case of the Philippines, in contrast to the examples of China, India, Japan, and the Republic of Korea, a weak scientific infrastructure had only a tangential relationship to the industrial structure.

It is not only the relationships between industrial and S&T organizations that are important, but also the internal workings of the organizations themselves. The industrial organizations in the West grew up with particular structures and characteristics as a part of an organic historical process. Attempts to transfer these Western organizational features wholesale do not necessarily succeed, and, when such transfers are made, the expected technological output may not replicate the success in a Western environment, as several studies going back to the early post-Second World War period have shown.[7] The studies here have not concentrated on the details of organizational social structures as a filter of technology; but, undoubtedly, these factors would have been important in the technological successes and failures of the different countries, as studies on the effectiveness of Japanese organizations demonstrate.[8]

The rural–urban relationship

As important a factor in industrialization and technology mastery as the relationship between the external global environment and the internal situation in a country is the rural–urban relationship within a country. It is this that affects the terms of trade and other relationships between town and countryside and between industry and agriculture, intimately colouring processes of technology absorption.

In China, the Republic of Korea, and Japan, extensive land reforms after the Second World War prepared the background for rural agricultural production and also a rural market for industrial products. In Korea and Japan, the high purchasing price of rice, several times above the world market price, enhanced this rural market pull. In Thailand, the growth of agriculture on an extensive basis developed the rural economy, although the intensity of agricultural development was less than in other countries. In India and the Philippines, although no significant agrarian reforms were made, the introduction of high-yielding varieties provided increased incomes for rural areas. The passage of technology to industry is helped greatly by enhancing the mutually rewarding relationships between town and country.

Yet, in a large country like India, even given a high degree of inequality, a large urban middle class estimated at 200 million[10] could still provide a viable market for some industrial consumer products. In the Philippines, the relatively small size of the urban middle class, combined with factors of dependence, has largely limited the level of industrialization.

Informal and formal sectors

In contrast to the formal S&T sector, which in some countries exists as an island unconnected with economic activities, there is also the informal knowledge system. This is also the knowledge practised by the Asian informal economic sector. This sector is responsive to immediate market demands and market openings, and uses knowledge opportunistically wherever it finds it. In the bazaars and small markets of Asia, the informal sector is thus continuously innovating.

Often, products from the formal sector are reprocessed and recycled into new artefacts. For example, an old thrown-away electric bulb and a discarded tin may be made into an oil lamp, a technological product found in many parts of South Asia. By similar innovative processes, new methods are discovered for making 30-year-old vehicles function, and new toys are made from junk. Real technological creativity occurs in this milieu and the formal S&T structure is largely tangential to this creativity.

The same informal creativity also exists at the higher ends of the technology spectrum, when repairmen and hobbyists carry out ceaseless experimentation and innovation in order to repair and modify difficult-to-obtain electronic products and parts, and to make novel customized products. To this same class of creative technology acquisition also belong those small enterprises, such as those found in Hong Kong, where new electronics-based products are constantly created on the basis of reverse engineering, product piracy, direct market pulls, and continuous inventiveness.[9]

It is also this type of informal creative culture which drove Silicon Valley in its heyday and which gave rise to the *original*[10] computer "hacker" culture, both good examples of technology creativity and mastery at the highest level. It is this dynamic creative relationship

363

that should exist between the formal S&T system and the economy. And, in the country which most successfully transferred technology, Japan, the boundary between the formal and the informal in organizations does not exist.[11] In this milieu, different technological cultures can blend into each other seamlessly, synergistically.[12] Here, a positive milieu for organizational technological creativity may have been created, in contrast to the more rigid arrangements seen elsewhere.

New generic technologies

The transformation from a pre-industrial economy to a technology-based agricultural and industrial economy which was attempted in many Asian countries – and which has been fully or partially achieved in several countries – occurred at a time when the developed countries themselves were undergoing a further profound technological transformation. If the beginnings of industrialization were associated with steam power and its later developments with electricity, the automobile engine, and chemical industries, the new "third-wave" technologies were those of information and biotechnology. They are more "generic" than any technology since the Industrial Revolution.[13] Within the next few decades, they and their products are expected to penetrate many aspects of the economy, the workplace, and the home, and hence any discussion of technological self-reliance in Asia today must take them into account.

Commentators have pointed out that, for the developing countries, information technology offered scope for "leap-frogging" in development.[14] The other new technology, biotechnology, could possibly have an equally pervasive impact in agriculture and medicine.[15]

With major chip manufacture and computer-related industries in Japan and the Republic of Korea, the East Asian region is one of the strongest manufacturing bases for information technology. This manufacturing also exists to varying degrees in India and China and in the next generation of would-be NICS, which include Thailand and, to some minimal extent, the Philippines. Some countries, such as India, have also been developing software as an export product, and this by 1991 had become a dynamic sector.[16]

Several of the country reports have taken note of the implications of the new technologies. The approaches can be roughly differentiated between the Republic of Korea and Japan, where strong organic links between S&T and industry exist, and India and China, where these links are weak. With product cycles in information technology counted in a few years and changes in biotechnology that are equally rapid, the technologies change faster than the five-year or ten-year planning cycles of governments. This implies that, for these new technologies, rigid S&T structures with weak links to industry are inappropriate and that more direct and fluid mechanisms for connecting the two are the ones that will succeed.

Social shaping of technology

The country studies have revealed aspects of the social effects and implications of technology acquisition. The studies supplement a large number of others done on Western societies, covering such aspects as the relationship of technology to society and how society shapes technology and vice versa,[17] as well as a smaller number of similar studies done on the Asian region. Could this combined knowledge be more directly applied to Asian societies in order to generate socially useful technologies, as well as to develop technologies that are associated with desirable social arrangements? With 40 years of experience, can one socially shape technologies in Asia in the future? In posing these questions the concept of self-reliance is crucial, since it implies the social shaping of the technology through local social and cultural needs and not through predominantly non-Asian ones.

This question is more important in that the newer technologies – information technology and biotechnology – are more socially responsive than the earlier ones. They can be "cut" and shaped socially in many more ways than the earlier technologies. Since the end of the field research for this study, a considerable number of research studies have been published on the social dimensions of the new technology. The new technologies, they indicate, are intimately coloured by the social assumptions of the societies that gave birth to them.[18] As a consequence, information technology[19] carries within it certain orientations and cultural assumptions arising from the way it has been socially constructed.[20] The variations in social shaping are seen, for example, in the very strong differences between nations in the characteritics of their computer-integrated manufacture (CIM) systems.[21] The social shaping is also seen in software, where cultural factors, including gender relations, influence the technology.[22] In a similar manner, social factors directly shape biotechnology.[23]

Conscious shaping of the technology

Recognition of this intimate relationship between biotechnology and information technology, on the one hand, and the social system on the other can allow for conscious social interventions in the technology, instead of the implicit intervention that usually happens, bringing a fresh dimension to the concept of Asian self-reliance. As the two technologies are very flexible to social and cultural pressures, the question then will be which culture's and which society's values will be mapped within these technologies as they unfold. A consequent task in self-reliance will then be to influence this most plastic of technologies so that it reflects the best social aspirations and knowledge systems of the cultures in Asia. Already, of course, the pre-industrial technologies of Asia have been blended with the old industrial technologies.[24] The problem will be consciously to guide this blending.

The Asian region is rich in non-European intellectual activity, including aspects of technology and the sciences. Some recent research indicates that these indigenous Asian aspects could become a useful adjunct to the development of new technology in Asia.

Thus, developing new biotechnology material requires access to a variety of useful genes. Plants that are unknown in developed countries have many uses that have been identified over the centuries by farmers and non-Western medical practitioners across Asia. Part of this knowledge is now being gathered by MNCs, the plants and their properties identified, and, later, the particular gene responsible for a desired property isolated and incorporated in a new genetically engineered plant.[25] Scouring Asia's past formal and local traditions for useful plants, and including these in genetically engineered products, would be a useful Asian contribution to the shaping of the new technology.

At the information technology software end, too, it seems that the diversity of Asian concepts of mental processing could possibly be used as a model on which software for information processing could be written. No single model of how the mind works, Western or non-Western, yet provides a complete picture of all aspects of mental processing and hence gives a perfect model to mimic in software. In several Asian civilizations, various systems of logic, epistemology, and psychology have been developed independent of the Western tradition,[26] and these could be written as software, in a way similar to that by which various competing and partial Western models of mental processing have been used as the basis of a wide array of information products. Some preliminary successes may have already been made in this direction, for example in language translation programmes that use the linguistics of the fifth-century BC grammarian Panini.[27]

The greatest impact on self-reliance could come from Asian answers to some of the troubling questions raised by the new technologies. The latter put doubt on some of the most cherished self-perceptions of humans. Biotechnology raises key questions about traditional concepts of what it is to be a living being, including what it is to be uniquely human, in the biological sense.[28] On the other hand, information technology, especially artificial intelligence (AI) that mimics human mental processes, raises questions about what it is to be uniquely human in a cultural sense. Real self-reliance will require searching answers to these questions within Asian cultural contexts.

Because of these key questions, debates on ethical and cultural issues are shaping both technologies. Thus, the release of biotechnological products into the atmosphere has been debated within a framework of its potential impact on other organisms.[29] And advances in medicine relating to, say, the onset of life and its termination have been hotly discussed and have influenced conventional medical technology. Developments in the new biotechnology stretch these questions very much further, raising fresh and very complicated ethical issues.[30] These discussions and controversies in the cultural and social sphere influence and continuously shape the new technology.[31]

However, the social and medical implications of biotechnology have as yet been largely discussed only in Western countries.[32] These debates have unfolded within a context that assumes Western cultural and social givens as universal; the imprint of the West's religious traditions, for example, is unconsciously brought in.[33] In Asian countries there has been little debate on these matters.[34] Yet, workers in the field have pointed out that Asian traditions could well give different answers to these questions,[35] as for example reflected in the Japanese response to definitions of clinical death.[36,37]

Advances in biotechnology, including gene therapy, could reshape and reformulate, among other things, life, death, health, and beauty.[38] The ethical as well as aesthetic criteria on which these are decided are deeply culture-bound and, if debated within the Asian region's different cultural traditions, would give different answers from those of the West. And this act of self-reliance would tend to give a different direction to technology.

Advanced information technology, especially AI-related technology, aims at cloning the partial behaviour of the mind. This again would raise profound questions for those parts of Asia which have strong cultural and religious traditions emphasizing the importance of the mind and mind culture. Asian inputs into debates on the ethics and nature of AI could also strongly influence the direction of information technology.

Existing agendas for shaping technology

The conscious social shaping of the new technologies, one should note, has already been attempted in a pan-European programme to develop "human-centred" technologies.[39] Here, a philosophy of "anthromopocentric" production systems has been adopted by the Forecasting and Assessment in Science and Technology (FAST) programme of research in Europe, which has an advisory relationship with the EC countries.[40]

The FAST programme studies production systems which emphasize desirable human qualities, and which take cultural differences into account. It has internalized the fact that in Europe there are many cultures and that technologies should accommodate this variety. It is expected that such anthropocentric technologies, because of their human scope, will be efficient and make European industry very competitive in the twenty-first century.[41]

Efforts similar to FAST, which take into account the social and cultural givens, are exercises in self-reliance that could be profitably emulated in the Asian region, whose cultures diverge from each other much more than in Europe. As the region increasingly adopted the new technologies, its local cultural bent would inevitably be stamped on them, often in implicit ways. The way to a more productive and socially relevant technological future is a strategy that uses the strengths of local cultural traditions, including their knowledge inputs.

A region which is fast emerging from the shadow of Europe, and which, earlier, was also home to some of the world's most vibrant civilizations, would find such an approach congenial. Making the new technologies socially responsive, in line with this extended view of self-reliance, would in addition increase both productivity and social well-being.

Concluding remarks

In this study we have noted some of the key aspects of the technology acquisition process in Asia. Several countries, especially the larger ones, have pursued broad-based strategies of technology acquisition. After nearly 40 years, the results are uneven. There are many notable successes, but there are also a number of shortcomings. In several countries, the potential of investment in technological acquisition has not been fully realized, and the technological system has not interacted fruitfully with the socio-economic and cultural systems.

Our study indicates that self-reliance in the mastery of technology requires cognition of, and action on, some key two-way relationships. These are the external geopolitical environment and internal social structures, the rural–urban dichotomy, S&T formal structures and industry, the informal and formal economy, and local and imported cultural and knowledge systems.

Some of the S&T strategies that have been built up over the last 40 years have helped develop these dialogical relationships, while some others have hindered them. As a result, over the decades many lessons have been learnt and internalized. These lessons could now be used for renewed efforts in S&T development.

Notes

1. Edgar Snow, *Edgar Snow's China: A Personal Account of the Chinese Revolution*, New York: Random House, 1981.
2. W. Joseph, "A Tragedy of Good Intentions – Post Mao Views of the 'Great Leap Forward'," *Modern China* 12 (1986), 419–457.
3. Richard Baum, ed., *China's Four Modernizations: The New Technological Resolution*, Boulder, Colo: Westview Press, 1980.
4. S. Arunachalam, "Why is Indian Science Mediocre?," *Science Today*, New Delhi, February 1979; S. Arunachalam and S. Markanday, "Science in the Middle Level Countries: A Bibliometric Analysis of Scientific Journals of Australia, Canada, India and Israel," *Journal of Information Science* (North Holland) 6 (1981), no. 3.
5. Personal communication from former Japanese Minister of Education, Michio Nagai.
6. Mark Crawford, "Japan's US R&D Role Widens, Begs Attention," *Science* 223 (1986); Eliot Marshall, "School Reforms Aim at Creativity," *Science* 235 (1986).
7. A.K. Rice, *Ahamadebad Experiment*, London: Tavistock, 1953.
8. Jon Sigurdson, ed., *Measuring the Dynamics of Technological Change*, London: Pinter Publishers, 1990.
9. Susantha Goonatilake, "Inventions and Developing Economies," *Impact of Science on Society*, Autumn 1987.
10. Esther Dyson, "Hacker's Rights," *Forbes*, 7 January 1991.
11. Shin-ichi Takezawa, "The Blue Collar Workers in Japanese Industry," *International Journal of Comparative Sociology* 10 (1969): 178–195.
12. See Fumio Kodama, "Japanese Innovation in Mechatronics Technology," in Sigurdson (note 8 above).
13. *Bio Technology: Economic and Wider Impacts*, Paris: OECD, 1989, pp. 52–55.
14. Amiya K. Bagchi, "The Differential Impact of New Technologies: A Framework for Analysis," ILO Working Papers no. 167, Vienna, 1987, p. 52.
15. UNCTAD, *Trade and Development Aspects and Implications of New and Emerging Technologies: The Case of Biotechnology*, Geneva: UNCTAD, 1991.
16. *The Times of India*, 18 June 1990.
17. Donald Mackenzie and Judy Wajcman, *The Social Shaping of Technology*, Philadelphia, Pa.: Open University Press, 1985.
18. "New Information Technologies and Development," *ATAS Bulletin* no. 3, New York: UN Center for Science and Technology for Development, p. 5.
19. C.A. Bowers, *The Cultural Dimensions of Educational Computing*, New York: Teachers College Press, 1988.

20. See note 19 above.
21. Karl H. Ebel, "Manning the Unmanned Factory," *International Labour Review* 128 (1989), no. 5; Karl H. Ebel, "Computer Integrated Manufacturing: The Social Dimension," Genera: ILO, 1990, p. 7.
22. Pamela S. Kramer and Sheila Lehman, "Mismeasuring Women: A Critique of Research on Computer Ability and Avoidance," *Signs: Journal of Women in Culture and Society* 16 (1990), no. 1: 171.
23. Erik Baark and Andrew Jamison, "Biotechnology and Culture: The Impact of Public Debates on Government Regulations in the US and Denmark," *Technology in Society* 12 (1990): 27–44.
24. A. S. Bhalla et al., *Blending of New and Traditional Technologies*, Dublin: Tycooly, 1984.
25. Celestous Juma, *The Gene Hunters: Biotechnology and the Scramble for Seeds*, Princeton, N.J.: Princeton University Press, 1989.
26. See the American journal *Philosophy East and West* (Honolulu) for numerous examples.
27. See, for example, Rick Briggs, "Knowledge Representation in Sanskrit and Artificial Intelligence," *AI Magazine* 6 (1985), no. 1; Rick Briggs, "Shastric Sanskrit as a Machine Translation Interlingua," paper presented at workshop on "Pannini and Artificial Intelligence," Indian Institute of Science, Bangalore, 1986.
28. Donnelley Strachan, "Medicine, Morality, and Culture: International Bioethics," Hastings Center Report, Special Supplement, July/August 1989.
29. Carol A. Hoffman, "Ecological Risks of Genetic Engineering of Crop Plants," *BioScience*, 40 (1990), no. 6.
30. Zbigniew Bankowski, "Ethics and Health," *World Health*, April 1989.
31. Dorothy Nelkin and Lawrence Tan Credi, *Dangerous Diagnostics: The Social Power of Biological Information*, New York: Basic Books, 1990.
32. Daniel Callahan, "Religion and the Secularization of Bioethics," *Hastings Center Report*, Special Supplement, July/August 1990.
33. James P. Wind, "What Can Religion Offer Bioethics?" *Hastings Center Report*, Special Supplement, July/August 1990.
34. See note 28 above.
35. Daniel Callahan and Courtney S. Campbell, "Theology, Religious Traditions, and Bioethics," *Hastings Center Report*, Special Supplement, July/August 1990; Courtney S. Campbell, "Religion and Moral Meaning in Bioethics," *Hastings Center Report*, Special Supplement, July/August 1990.
36. See note 28 above.
37. Kajikawa Kin-ichiro, "A New Field Emerges," *Hastings Center Report* 19 (1989), no. 155.
38. Marge Beree, "The Perfection of Offspring," *New Scientist*, 14 July 1990, pp. 58–59; Jean Bethke Elshtain, "Reproduction Ethics," *Utne Recorder*, Iss. 44, March 1991.
39. Peter Brodner, "Towards the Anthropocentric Factory," Tokyo International Workshop on Industrial Culture and Human Centered Systems, Tokyo Kezai University, 1990.
40. Werner Wobbe, "Anthropocentric Production Systems in the Context of European Integration," Tokyo International Workshop on Industrial Culture and Human Centered Systems, Tokyo Kezai University, 1990.
41. Felix Rauner, "Changing World of Industrial Culture," Tokyo International Workshop on Industrial Culture and Human Centered Systems, Tokyo Kezai University, 1990.

Contributors

Editors

Saneh Chamarik, Former Director, Thai Khadi Research Institute, Thammasat University; Local Development Institute, c/o Department of Medical Science, Bangkok
Susantha Goonatilake, Visiting Scholar, New School for Social Research, New York

Project coordinator

Saneh Chamarik

Introduction

Susantha Goonatilake

India

V.R. Panchamukhi (Joint Country Coordinator), Director, Research and Information System for the Non-aligned and Other Developing Countries
S. Chakravarty (Joint Country Co-ordinator), Chairman, Economic Advisory Council
Other contributors: Subhash Dhigra, Vijay Kelkar, Kalyan M. Raipuria, Jairam Ramesh, S. Setlur, Nagesh Kumar, Biswajit Dhar, Kishore Jethanandani, and P.R. Bhatt

China

Wang Huijiong (Country Coordinator), Executive Director, Development Research Centre, State Council
Li Poxi

Republic of Korea

Tai-Wan Kwon (Country Coordinator), President, Korea Food Research Institute
Other contributors: Chong Ouk Lee, Hyung Bae, Yong Sun Kim, Young Kun Shim, and Chun Su Kim

Thailand

Thamrong Prempridi (Country Coordinator), Head, Department of Civil Engineering, Rangsit University
Other contributors: Mongkol Danhanin, Vishan Poopath, Boriboon Wongsarnsri, Pirom Chanthavorn, Bharata Kunjara, Bundit Jarimopas, Charnchai Limpiyakorn, and Paibul Chainil

Philippines

Roger Posadas (Country Coordinator), Dean, College of Science, University of the Philippines

Japan

T. Shishido (Country Coordinator), Former Vice-Rector, International University of Japan; Chairman, Research Institute of Construction and Economy
Other contributors: T. Hayashi, I. Inukai, S. Kimura, H. Uchida, Y. Okada, and H. Tada

The lessons from Asia

Susantha Goonatilake